I0428117

Fish

WITHDRAWN
UTSA LIBRARIES

Fish

Elizabeth R. DeSombre
J. Samuel Barkin

polity

Copyright © Elizabeth DeSombre & Jeffrey Samuel Barkin 2011

The right of Elizabeth DeSombre & Jeffrey Samuel Barkin to be identified as Author of this Work has been asserted in accordance with the UK Copyright, Designs and Patents Act 1988.

First published in 2011 by Polity Press

Polity Press
65 Bridge Street
Cambridge CB2 1UR, UK

Polity Press
350 Main Street
Malden, MA 02148, USA

All rights reserved. Except for the quotation of short passages for the purpose of criticism and review, no part of this publication may be reproduced, stored in a retrieval system, or transmitted, in any form or by any means, electronic, mechanical, photocopying, recording or otherwise, without the prior permission of the publisher.

ISBN-13: 978-0-7456-5019-7
ISBN-13: 978-0-7456-5020-3(pb)

A catalogue record for this book is available from the British Library.

Typeset in 10.25 on 13 pt Scala
by Servis Filmsetting Ltd, Stockport, Cheshire
Printed and bound in Great Britain by MPG Books Group Limited, Bodmin, Cornwall

The publisher has used its best endeavours to ensure that the URLs for external websites referred to in this book are correct and active at the time of going to press. However, the publisher has no responsibility for the websites and can make no guarantee that a site will remain live or that the content is or will remain appropriate.

Every effort has been made to trace all copyright holders, but if any have been inadvertently overlooked the publisher will be pleased to include any necessary credits in any subsequent reprint or edition.

For further information on Polity, visit our website: www.politybooks.com

Library
University of Texas
at San Antonio

Contents

Abbreviations

ASEAN	Association of Southeast Asian Nations
CAFO	Confined Animal Feeding Operation
CCAMLR	Commission for the Conservation of Antarctic Marine Living Resources
CITES	Convention on International Trade in Endangered Species of Wild Fauna and Flora
CPR	Common Pool Resource
CPUE	Catch Per Unit Effort
EEZ	Exclusive Economic Zone
ENSO	El Niño Southern Oscillation
EU	European Union
FAO	United Nations Food and Agriculture Organization
GDP	Gross Domestic Product
GPS	Global Positioning System
ICCAT	International Convention for the Conservation of Atlantic Tunas
IFQ	Individual Fishing Quota
ISA	Infectious Salmon Anemia
ITQ	Individual Transferable Quota
IUCN	World Conservation Union
IUU	Illegal, Unreported, and Unregulated Fishing
IWC	International Whaling Commission
MICRA	Mississippi Interstate Cooperative Resource Association
MPA	Marine Protected Area

MSC	Marine Stewardship Council
NAFO	Northwest Atlantic Fisheries Organization
NGO	Non-Governmental Organization
NOAA	National Oceanic and Atmospheric Administration
OECD	Organization for Economic Cooperation and Development
OSPAR	Convention for the Protection of the Marine Environment of the North East Atlantic
PCB	Polychlorinated Biphenyls
RFMO	Regional Fisheries Management Organization
TAC	Total allowable catch
TED	Turtle Excluder Device
UK	United Kingdom
UNCLOS	United Nations Convention on the Law of the Sea
UNDP	United Nations Development Programme
UNEP	United Nations Environment Programme
USA	United States of America
VMS	Vessel Monitoring System

CHAPTER ONE

Introduction

You sit down at a restaurant, considering what fish to order, and choose a nice panko-crusted Chilean sea bass.

Long before the fish reached the restaurant menu, a set of marketing executives had decided that "Chilean sea bass" sounded more appealing than Patagonian toothfish, the fish's true name, despite the fact that the species is in no way related to an actual sea bass (which is a type of grouper) and is only occasionally caught in Chilean waters.

It was, nevertheless, first commercially marketed internationally after Chilean President Augusto Pinochet opened up Chilean waters to foreign fishing trawlers in the 1980s. The newly exported fish quickly caught on in United States (and then Japanese and European) markets and by the 1990s was the new hot menu item in restaurants, increasing the price paid to fishers for capturing the fish, and dramatically increasing catches worldwide. *Bon Apétit* magazine named it the "dish of the year" in 2001. One of the other reasons for the fish's newfound prominence on restaurant menus was the depletion of stocks of other fish, such as cod, or even white sea bass (which is an actual sea bass).

Before the fish on your restaurant plate was caught, it lived in cold waters near Antarctica. If it hadn't been caught it might have lived to an age of 50 years and could have reached a length of 2.3 meters (7.5 feet). It grew slowly, reaching reproductive maturity at about 90 to 100 centimeters in length, at between about nine and twelve years of age, so if it was caught before then it would not yet have had a chance to reproduce. The combination of long life and slow maturation makes it difficult to tell if a species is being overfished. Because it is difficult to regulate catches before there is clear evidence of overfishing, Patagonian toothfish was a strong candidate for fast overexploitation once the species became a commercial success.

If your fish was caught by trawling, a fishing vessel dragged a net over the sea floor to take it in, damaging the sea floor ecosystem and capturing other fish and non-fish species that were probably discarded and subsequently died. This bycatch accounts for as much as one-quarter of what is caught in the world's fisheries. More likely, the fish was caught by a long-line, a set of up to hundreds or thousands of baited hooks (fewer if it was caught by a vessel following international regulations) on one line, reeled out behind fishing vessels and weighted to sink to between 500 and 2,500 meters deep. These longlines often catch seabirds (especially the endangered albatross) or non-target fish, most of which die before they can be released.

The vessel used to catch the fish was a factory fishing vessel, large and well-equipped to be able to stay at sea for months at a time. It can process catches on board, so after the toothfish are caught their heads and tails are removed, they are gutted, and flash-frozen for storage in the hold. Such vessels can hold 300 tonnes of fish before returning to port. The technological revolution allows the fishers to use GPS to find their way to and from the places they set their lines, and sonar to map the seafloor or locate schools of fish, to help the fishers identify likely areas to find the toothfish.

The fishers who caught the fish should have been following the rules set out by the Commission for the Conservation of Antarctic Marine Living Resources (CCAMLR), but odds are that they weren't. Even now, at least twice (and by some estimates five times) the amount of legal catch of toothfish is caught illegally. In the legal gray area are those vessel-owners who register their fishing vessels in countries that are not CCAMLR members. By international law ships are bound by the domestic and international rules their registry state has adopted, so a country that is not a CCAMLR member is not required to follow its catch rules.

Recent efforts by CCAMLR to better protect the stock from unregulated fishing, however, mean that in order to land fish in member countries (including the major toothfish markets of the United States, Japan, and the European Union) the fishers should have to demonstrate that their catches were done in a way consistent with international rules. But it also turns out to be reasonably easy to avoid even that level of regulation. When the rules were first imposed, ships reported catches in a statistically improbable distribution outside of the regulatory area, suggesting that they were perhaps not honest about where their catches took place. More recent efforts to require satellite tracking on ships to verify the locations of their catches have increased the ability of international organizations to determine whether fish were caught legally. But in some cases, the tracking systems report information only to the registry state, thereby avoiding international oversight, and some states may benefit from overlooking the illegal behavior of fishing vessels they register. This tracking process does make the fish sold in CCAMLR member states more likely to be caught within the international rules, but it is still difficult to know. And fish that were caught outside of the regulatory process are indeed able to be sold and consumed, if not in CCAMLR member states, then elsewhere.

If you had carried around your *Seafood Watch* pocket guide to sustainable fish consumption, from the Monterey Bay Aquarium, you would have found Chilean sea bass on the list of fish to avoid. Likewise, if you had checked with the

Environmental Defense Fund you would have discovered a health advisory for Chilean sea bass; like other fish reasonably high on the food chain, it biomagnifies the mercury contained in the smaller fish it consumes. On the other hand, if the fish came from the bottom-set longline fishery off the island of South Georgia, it would have been certified by the Marine Stewardship Council as originating from a sustainably managed fishery.

All in all, you're wondering whether it might have been simpler to order the pasta instead.

State of the World's Fisheries

Fish are a key food source for much of the world's population, providing 15 percent of the animal protein consumed globally; nearly one-quarter of the world's people receive more than 20 percent of their animal protein from fish, and in some places in Asia and Africa (as well as in small island developing states) that number exceeds 50 percent.[1] But the ability of the resource to sustain this level of consumption is in serious doubt. Increasingly intense and environmentally degrading commercial fishing methods have called the long-term sustainability of fish stocks into question. One particularly alarmist estimate predicted that global commercial fisheries will completely collapse by 2048.[2]

Not all estimates are quite that pessimistic, but it is clear that the health of ocean fisheries is not strong. The long history of industrial fishing has had a disastrous effect on global fish stocks. The United Nations Food and Agriculture Organization (FAO) estimates that 80 percent of world fish stocks can withstand no increase and may not even be able to sustain the level of fishing currently experienced. At least one-quarter of the world's ocean fish stocks are considered overexploited or depleted, half are fully exploited (permitting no additional fishing growth), with less than one-quarter capable of sustaining any growth in catches.[3] In short, we are, on the whole, not managing the world's fisheries sustainably.

The fish stocks in greatest danger are high-seas, highly migratory fish that are top predators. These fish stocks, which include tuna and swordfish, have declined by up to 90 percent since industrial exploitation began in the 1950s.[4] Cod in the Northwest Atlantic, and salmon, pollock, and halibut in the Atlantic as a whole, are nearly commercially extinct, meaning that they cannot support a commercial fishing industry. The declines of these major predators in turn have effects throughout the ecosystem that often decrease, rather than increase, diversity and abundance of other fish species used for human consumption. Meanwhile, fishing operations move down the food chain and deplete the fish that those top predators had originally consumed.

The loss of biodiversity in the ocean that can be attributed to commercial overfishing leads to a disruption of the productivity of ocean ecosystems more generally, and therefore also disrupts opportunities for future commercial fishing. These changes also affect other ecosystem services such as water quality. Other human-caused problems, like pollution and climate change, have contributed and will increasingly contribute to the stress on fisheries.

The increasing intensity of fishing practices, and expansion and increasing technological sophistication of the fishing industry more generally, as discussed in detail in Chapter 2, have had a devastating effect on the number of fish available for capture. The broader context of who has access to fish and what has accounted for that level and type of access, provides essential background for understanding the context of increased fishing intensity and depletion of fish stocks.

Diagnosis of the Problem

Many factors have contributed to current and projected problems in the world's fisheries. As is true with other commodities, industrialization has led to more efficient and larger-scale resource extraction and transport. At the same time, a growing world population with rising income has turned increasingly to fish to meet protein needs, as well as needs for healthy oils. Interrelated increases in supply and demand have resulted in the current crisis in ocean fisheries.

There are some ways in which fish are different from other types of resources. With many global commodities, too much control by self-interested profit-seeking actors may generate monopolies or cartels, preventing access to the commodities by those who rely on them, as is the case with industrialized food production or petroleum. The opposite is true with ocean fish. Too many people have access, and there is too little control over what any of them can do.

Issue Structure
The reason for the difference between fisheries and most other natural resources is the issue structure of the resource. Two elements of this difference are key: the ownership of fisheries, and the way that fisheries resources reproduce themselves. These elements of issue structure are exacerbated by uncertainty about many aspects of fish biology and ecology, and the effects of exploitation on fish populations.

Unlike most resources, fish are generally unowned. A farm or an oil well is generally owned by a person or company (or at minimum someone has the exclusive right to farm or pump from a particular patch of land). That person or economic entity thus has an interest in using the resource in a way that maximizes its long-term potential, because if it is abused, the owner suffers most from the effects. And when ownership is not clear, bad things often happen. When people farm land to which they do not hold clear title, they are less likely to maintain the land well. When countries pump oil that is not clearly theirs, international incidents, and sometimes major wars (such as the Iraqi invasion of Kuwait) can happen.

But it is rare for a fisher to own a particular fish until it is caught. This means that fisheries work on a first-come, first-served basis in a way that is not generally the case with other resources. The absence of clear ownership means that individual fishers do not have a clear incentive to restrain their fishing today for the sake of the health of the fishery in future years, because if they do not catch a fish, somebody else probably will. (Note that aquaculture – fish farming – does not work this way, making it a very different industry from the capture of fish in the wild, as is discussed in Chapter 5.) An area in which a resource is available on a first-come, first-served basis is called a commons, and the resource itself is called a common pool resource (CPR). Most bodies of water are examples of commons, and the ocean is a global commons. Fisheries are a classic CPR.

Common pool resources have two key characteristics. They are subtractable (or what economists would call "rival"), meaning that use of the resource by one actor diminishes its usefulness for others. Fish taken from the ocean can no longer be caught by others, and they are no longer there to reproduce and thereby create new fish to be caught by others. And they are nonexcludable – it is difficult, both physically and, often, legally, to prevent access to them.

These structural characteristics create a set of incentives that make fishery problems particularly difficult to address. Fishery problems create tensions between collective and individual incentives, and between long-run and short-run incentives. On the one hand, because those who catch fish are, in part, harmed by their own fishing activity that decreases the abundance of fish, the very actors engaged in depleting the fishery are those who would benefit if the resource could be successfully protected. Collectively, fishers will benefit from successful management of fisheries, because it will enable them to keep fishing for the long-term.

But the temptation to free-ride on others' protection of the resource is clear. As Garrett Hardin pointed out in his important overview of the "Tragedy of the Commons,"[5] those who use a common pool resource gain all the positive utility of their action (in this case, the fish that has been caught) but only a portion of the negative utility (its absence from the ecosystem), which is shared with all other users. In the short run, therefore, the decision is often weighted toward the individual advantages to the user of the resource, who bears only a fraction of the downside of his or her resource use. The best option for a given fisher is to take a lot of fish while everyone *else* in the system refrains from overfishing. The one who does not exercise restraint can take advantage of the increased supply of fish. The problem, of course, is that if everyone acts that way, there is no conservation of the resource. Moreover, even if fishers can be persuaded that collective action to restrain fishing behavior is in everyone's interest, the fear that others will not go along with the action may undermine an individual's willingness to do so. In game theory the "sucker's payoff" refers to the situation in which you act for the collective good and others do not; the restraint you exhibit might have been useful if everyone else did as well, but instead you lose twice: by limiting your fishing behavior in the first place, and because the fish stock will not be saved so you will not reap the long-term benefit of your conservation action.

And because of the subtractability of the resource, efforts to limit fishing behavior to protect the resource are even more difficult. In some cooperative enterprises, free riders – those who do not contribute their fair share to the endeavor – might not undermine the cooperative outcome. The outcome in these cases is called a public good, which is different from a CPR in that it is not subtractable. A person who watches public television without contributing during the annual pledge drive will not help keep the channel on the air, but does not diminish it by her viewing activity. Unlike the provision of public goods, in which a small committed group of actors can choose to take action without the participation of all involved and thereby address a global problem, a common pool resource cannot be successfully protected by a sub-group of users. Anyone who does not join a collective fishery management agreement or consent to refrain from overfishing can actively undermine the ability of those who are cooperating to address the problem.

Many of the world's fishery resources are entirely contained within countries, or within their exclusive economic zones (EEZs). These are areas of ocean up to 200 nautical miles from shore where countries can claim exclusive rights to manage economic activity. From the perspective of fisheries management, fish within EEZs are owned by the country whose EEZ the fish are in. Countries should have an incentive to manage these resources well, but still face the problem that, from the perspective of fishers, fish are CPRs. They are not owned by the individual fisher, and other fishers can still access that fish; once someone else has caught it, it is no longer available for you to catch. For that reason, even within an EEZ, most fishers behave as though the fishery is a common pool resource.

Other fishery resources, including some of the most valuable fish (such as Patagonian toothfish) either live in international waters, or swim between different EEZs or between EEZs and international waters. These create a CPR problem at the international level. Not only are these fish CPRs from the perspective of individual fishers, but they are CPRs from the perspective of governments as well. A fishery in international waters is first-come, first-served from the perspective of a government that wants to promote its fishing industry; if fishers from that country do not catch those fish, fishers from somewhere else will.

The time horizons of fishers and politicians add to these difficulties. Protecting fisheries requires taking costly or difficult action in the short term that will likely have benefits in the future, if everyone actually contributes. That approach is difficult to embrace. Individuals have good reasons to value the present over the future: food, or profits, now can be used for current needs or to plan for future ones. A fish left in the ocean may help create more fish for the future, but a fish caught now can be eaten or sold now. Politicians, as well, have clear time horizons: causing suffering (by restricting the ability of people to fish, for instance) that will not have paid off with beneficial results before the next election is a recipe for political defeat. It should not therefore be surprising that politicians have been reluctant to take sufficient action to require a reduction of fishing effort by their constituents.

The second key way in which the issue structure of fisheries is different from that of most other natural resources is the supply of fish. Many natural resources, such as metals or oil, are non-renewable – there is a certain amount out there, and once it is used it is not replaced. Others are renewable, but the regrowth is directly assisted by humans. Agricultural crops, for example, tend to be replanted by the same people who harvested the previous lot. Fish are renewable, but (with the exception of aquaculture, yet another reason why it is fundamentally different from capture fisheries) we count on them to renew themselves without our assistance. Fishers do not hold eggs in reserve and restock the ocean when they are done fishing.

There is thus a delicate trade-off between the economic activity of catching fish, and the need to leave enough fish behind to replace the ones caught. If too many are caught, there will be fewer next year. If far too many are caught, there may be none next year. A tricky element of the trade-off is that fish stocks often need a certain density in order to maintain themselves, in order to find mates efficiently and prevent encroachment on their ecological niche by other species. So it is not the case that one has only to leave a few fish to reproduce, even if they can often generate thousands or millions of eggs per female. The population level necessary for a stock to maintain itself may be reasonably high. Furthermore, fish populations often decline rapidly when fished beyond a sustainable level, meaning that it is easy to fish beyond this level and have the stock collapse without warning.

The irony here is that renewable resources such as fish can be easier to use up than non-renewable resources. A non-renewable resource will gradually get scarcer, and therefore more expensive to extract. As it gets more expensive, people should start using less of it, reducing demand pressure on what remains. The market for non-renewable resources is self-correcting in this way. But as a renewable resource, the market for fish does not self-correct in the same way. The market often does not send signals to fishers to catch fewer of a species as it gets closer to the line of overexploitation, and as a result more help is needed from regulators to correct overexploitation that undermines the ability of the resource to renew itself.

All of these factors, and in particular the tradeoff between current and future benefits and the question of the renewability of the resource, are even more difficult to manage in the context of uncertainty about the state of the fish stock or the behavior of fishers or regulators.

Uncertainty
Uncertainty is a key reason that global fisheries are particularly difficult to manage well. There is uncertainty about almost

every aspect of fisheries, about such things as population estimates, reproductive cycles, ecosystem characteristics, and human behavior. This uncertainty colors the behavior of individual fishers, of countries making regulatory and resource decisions, and of international organizations attempting to prevent overfishing.

There is, first, a lot of basic information about fish about which we are uncertain. At any given point in time no one actually knows how many fish of a given species there are in a particular area of the ocean, and it is often unclear whether those numbers are increasing or decreasing. At some points in recent history that uncertainty has been central to decisionmaking difficulties, particularly about international management of fisheries. Population estimates are difficult for any species, but much harder when the species cannot be directly observed, as is the case under water.

Population estimates in fisheries have been done in part by examining catch statistics and how the difficulty of catching fish has varied over time or season (with the understanding that as it becomes harder to catch fish, there must be fewer of them to catch). Because technology has changed over time (thereby allowing people to catch fish more efficiently) these estimates are usually calculated as Catch Per Unit Effort (CPUE). And for a variety of reasons, these estimates may not be reliable. In the first place, many fishing techniques produce extensive bycatch, the harvesting of unintended fish that may even be discarded. Gaining an accurate assessment, across all fisheries, of how much of what species is being caught is therefore extremely difficult, since fish that are bycatch in one fishing operation may be the target species in another.

Second, there is a lot of fishing that is done outside of the regulatory process. Some of that is decidedly illegal: fishers catch more fish than they are allowed and document only their legal catches. Some of it exists in a legal gray area: fishing vessels are registered in locations that are not members of international regulatory bodies; they are thereby not bound by rules made by those organizations, and not obligated to report information about their catches. And in some instances a lack of reporting may be in part motivated by the effort to conceal information that might lead to the ending of a fishing season. These preceding factors together constitute what some refer to as IUU fishing: illegal, unreported, and unregulated. In addition, a large number of small-scale or recreational fishers catch fish in ways that may be unregulated and therefore often unrecorded.

CPUE is not the only way to estimate abundance in a fishery. Some tagging programs try to count individual fish. Biomass estimates attempt to calculate the number or weight of fish in an area. To these are added other known details about characteristics of the fish species and its likely reproductive processes in complex (and controversial) equations used to calculate the likely number of fish of a given species in a specific area. An additional aspect of complexity is that most of these models are set up to deal with linear effects, and fish population dynamics are often non-linear, with major crashes in population possible.

We also need to know the life cycle of a fish species. At what age does it reach reproductive maturity? How many fry (babies) does it have, and how many of them will survive to adulthood? In what seasons, locations, and conditions does it reproduce? These characteristics can be learned, but they often are not until after a species has been overfished when too many juveniles are caught before they have had an opportunity to reproduce. For example, Greenland halibut, also known as Greenland turbot, is slow-growing; females do not reach reproductive maturity until nine years of age. That means that any fish caught before the age of nine will not have had a chance to reproduce, and the possibility that fish will be caught before that time is high. Although such demographic characteristics are knowable (even if research may be required to ascertain the information), those who fish for the species may not initially be aware of them and relevant regulatory processes may not initially take them into consideration.

The location and migratory behaviors of fish are also key sources of uncertainty. Atlantic bluefin tuna exist in both the western Atlantic, spawning in the Gulf of Mexico, and the eastern Atlantic, spawning in the Mediterranean. Until recently it was unclear the extent to which fish in these areas were part of the same stock, and the International Commission for the Conservation of Atlantic Tunas (ICCAT) regulates them as separate stocks, with separate quota allocations for those fishing them. But recent tagging studies demonstrate that individual fish regularly migrate across the Atlantic and that the intermingling level of the two stocks is actually quite high. Moreover, the lack of knowledge about migratory patterns may have led to an overestimation of the western populations. Although there are political reasons behind the separation of species into two stocks for regulatory purposes, the underlying belief that they did not intermingle made it appear reasonable to do so. New understanding of basic species behavior now suggests major problems with that approach.

Also important, and a possible source of uncertainty, are ecosystem characteristics that affect the lives and lifecycles of fish. Fish exist as part of a complex food web: how abundant or healthy the food source of a given species is can be important to its well-being, as can the abundance of the species that prey on the fish in question. And as some species are overfished or otherwise disappear from an ecosystem, others may fill ecological niches previously held by those overfished species, thereby changing the composition of species and the ability of a particular fish or species to survive.

Many (although perhaps not all) of the biological or population characteristics of a fish species are knowable, even if they may require research to determine. But human behavior is also uncertain and has a major effect on fish populations. How people respond to decreasing fish stocks, or to collective action or regulation attempting to protect fisheries, will have the most important impact on the health of fisheries, and that human behavior is quite difficult to predict. When stocks are declining, will fishers work together to restrain their fishing to allow the stocks to recover? Or will they redouble their effort to catch the remaining fish before someone else does? Will they accurately report their catch statistics in an effort to aid modeling of population dynamics or the enforcement of rules, or will they hide catches for fear of contributing to the end of the fishing season or increased regulatory stringency? To add to the complexity, how individual fishers react to conditions or regulations may be influenced by, and in turn influence, how others respond.

Ocean Pollution and Other Ecosystem Effects
Other things that change the characteristics of the ecosystem, like pollution or climate change, can be additional sources of uncertainty. It may not be clear how much of what types of pollutants are in an area, nor at what level or in what ways those pollutants affect the fish. Some of the major types of pollutants affecting ocean ecosystems and the fish themselves (and the health of human and other animal populations that eat them) are PCBs and heavy metals, particularly mercury, as well as nitrogen, oil pollution, and plastic. Other changes to an ecosystem, from dams or climate change, can also have devastating impacts on fisheries.

Polychlorinated Biphenyls (PCBs) are persistent organic pollutants used in a variety of electronics functions before they were phased out by most countries in the late twentieth or early twenty-first century because of their toxic effects. Fish contaminated with PCBs experience decreased reproductive success and changes in their immune systems. There are ecosystem effects on other fish, birds and mammals that consume PBC-contaminated fish. The effects on humans are especially troubling, and include liver damage and other immune system responses, as well as increased cancer risk.

Mercury is found in at least trace amounts in almost every fish species. It mostly originates from people burning coal on land for energy. Mercury exists naturally in coal and when it is burned the mercury is released into the air; it then falls into the ocean. Mercury contamination is a particular problem in fish species with long lifespans or that are reasonably high on the food chain, because both these characteristics give the fish a greater opportunity to ingest mercury. Species with particularly high mercury contamination include orange roughy, marlin, grouper, swordfish, tilefish, King mackeral, and shark, although other predator fish, like tuna, halibut, and Patagonian toothfish also can have quite high levels. While much of the discussion of mercury contamination in fish focuses on its effect in people who consume the fish, mercury causes direct problems for the fish themselves, causing behavioral changes and decreasing reproductive effectiveness.

Other types of pollution can impact fish and their habitats. Both individual events like major oil spills and day-to-day consistent pollution can devastate fish populations. The Gulf of Mexico provides two recent examples of both. The major spill from the failed deep-water well off the coast of Louisiana in the summer of 2010 will cause major damage to the region's fisheries, but we will not know for years how bad, or how long-lasting, it will be (and we may never know – the effects of the spill can be difficult to disentangle from the effects of other pollution and habitat loss).

Another major source of pollution in the Gulf of Mexico (as well as in other regions) has come from nitrogen runoff that has for decades created a "dead zone" between the Mississippi River delta and the coast of Texas, in which few exploitable fish stocks survive. Nitrogen is a primary component of fertilizer in industrial agriculture; the excess beyond what the plants or soils take up runs off the land in great quantities and into the water supply. In this region it enters the Mississippi river from farming practices in the American Midwest, as far away as from Minnesota, and are carried from there to the Gulf. At over 20,000 square kilometers, the dead zone is the size of Wales, and almost as big as Massachusetts or Belgium.

Other types of pollution have the potential to affect the health of fish stocks. Increasing amounts of plastic pollution have accumulated in the oceans, with some areas, like the North Pacific Ocean, now experiencing large gyres of plastic and chemical sludge as big as some continents. Plastic in the ocean can harm fish in a number of ways. Marine species can become entangled in or injured by large pieces of plastic. Fish can ingest small pieces of plastic, leading to the inability to gain proper nutrition from other sources or to other digestive problems. The toxic nature of some of the plastic, or of associated substances in the water that can cling to the waterborne plastic, can have health impacts on the marine species as well as for those who consume them.

More broadly, habitat loss is a threat to global fisheries. We may not know if the decline of a particular fish stock is the result of overfishing or habitat loss. An example here is salmon in the Pacific Northwest, which are threatened, on one hand, by industrial fishing, and on the other by dams and pollution in rivers where they spawn. Habitat loss can also include damage to coastal structures like mangrove forests and coral reefs, the acidification of freshwater lakes, as well as any other form of encroachment on lakes, rivers, and oceans by human development. Habitat loss can decrease the future population of fish.

Climate change in particular is likely to have a dramatic effect on fisheries, both from changing the ocean temperature and currents and thereby the habitability of certain ocean areas for certain species, and from other effects, like increasing acidification of the oceans from effects of increased ocean CO_2 uptake. The extent of climate change and its effects on oceans and fish is a source of extreme uncertainty, but some changes along these lines have already been detected. There is evidence of changes in fish feeding behavior and migration patterns that are likely to increase as climate change alters the temperature and chemical composition of the oceans. Alarmingly, the global quantity of phytoplankton, the plant life that serves as the bottom of the marine food chain and upon which all fish species ultimately depend, has decreased by 40 percent since 1950, a change that is correlated with, and almost certainly attributable to, the sea surface temperature change that comes from climate change.[6] The effect from such changes may be devastating to global fisheries.

The Political Economy of Fisheries

The interrelationship of politics and economics plays a key role in the depletion of international fisheries. Fish are bought and sold, both in global and in local markets. In principle, markets are efficient ways to allocate goods and services based on preferences. But because fish are common pool resources, the markets that underpin trade in them are structurally problematic, for a number of reasons. The first is a lack of clear property rights. Central to the idea of a market is that people buy and sell things that they own. But no one owns fish until they are caught, which is part of the reason there is overfishing: if you leave the fish in the water you have no claim to it, so it is safer to own it by taking it out of the water.

And a fish in the water is free – a fisher does not have to pay anyone for it. The value to society of that fish in the water is not taken into account, and the full economic costs of fishing are not accounted for in the market. Economists refer to this problem as a negative externality. Externalities are the unintended, and unpriced, consequences of economic activity. If you buy a fish, you pay for the fish, and perhaps the price of time and fuel and equipment that went to catching it (with the caveat that, because of subsidies, you may not even be paying that full price). But you do not pay for the cost of the future reproduction the fish will no longer accomplish, or the cost of ecosystem disruption from overfishing, or the other fish or bird species that were caught incidental to the catching of that fish. And you almost certainly do not pay for the harm your fishing technologies (as described in Chapter 2) impose on the ecosystem. The true cost of fishing is therefore much higher than the price anyone pays as a fish is bought or sold.

Some countries have developed systems for charging fishers for the right to fish, through the sale of licenses and quotas, which helps to ameliorate this problem. But it is not a perfect solution. A perfectly functioning market relies on complete and shared information, and for this perfect market to happen you need to know what the fish are worth. Regulatory systems that charge fishers for the right to fish can at best provide only a rough guess at the value of a fish in the wild. The uncertainty described above means that there is imperfect information about the future availability of fish, and that makes valuing existing fish as they are bought or sold difficult. Even in these approaches, the market does not price externalities on its own; it requires political intervention to account (if only imperfectly) for those externalities.

States "own" fish in their internal waters (lakes and rivers) and national waters (which – as explained in Chapter 4 – expanded dramatically in the 1980s with the creation of exclusive economic zones, or EEZs). But state ownership of fish is not the same thing as standard property rights; even when states are able to regulate, the fish species may be subject to tragedies of the commons as long as individual users do not have specific rights to certain allocations of fish to the exclusion of others. When governments manage these fish stocks, if they are able to do so, they do it on behalf of national-level goals, which may not involve the interest of some subset of their populations, and almost certainly do not involve the interests of the ecosystems in which the fish exist.

In addition, in a perfectly functioning market, fishers who are not able to make enough money fishing to make it worthwhile to continue to do so (because fish stocks are not sufficient to sustain the level of fishing activity, given what the fish can be sold for) will choose to leave the profession. But several factors prevent the market as it exists from providing that signal to fishers. First, governments often heavily subsidize fishing to encourage fishers to enter, or remain in, the profession. Governments give grants or loans at below-market rates that enable fishers to purchase larger vessels with advanced technology for increasingly efficient fishing. They subsidize fuel, or guarantee insurance, or otherwise contribute to the operating costs of fishers. Governments purchase access to foreign fishing grounds for their fishers. They provide fishing infrastructure, such as creating or maintaining the ports that fishers use without having to pay the full costs. And in some cases they provide price supports or income supports to ensure that fishers are able to continue to make a respectable living while fishing, even if the market would not provide sufficient income. Governments may do so to encourage employment or to help provide food for their populations, but these subsidies mean that fishers are not bearing the true costs of their fishing activities. Up to one-quarter of the income of the global fishing industry (and by some estimates, an even greater percentage) comes from government subsidies.[7] Without these subsidies, far fewer people, and far fewer vessels, would be in the industry, leading to far less pressure on global fish stocks.

In the face of rapidly increasing fishing effort over the course of the past half-century or so, governments have undertaken increasingly active efforts to manage fisheries to prevent overfishing and depletion of stocks. Some of these efforts have been at least partially successful, but many regulatory efforts to protect fisheries have been insufficient or counterproductive. With experience, fisheries regulators are learning more about how to create rules that are effective, and how to design regulations that work with rather than against the structure of the issue. An ingredient that is too often missing, however, is the political will to put stringent rules into effect. What is needed most at this point in time is an increase in the political will to manage the world's fisheries sustainably.

Plan of the Book

The next chapter sets the groundwork for this book, discussing the development of the global fishing industry, with a particular focus on fish in the oceans (marine fish, as opposed to freshwater fish). It looks at the growth of the industry and its recent plateau and subsequent decline as the limits of the ocean to provide us with fish were reached in the 1980s and 1990s. It examines the development of the fishing technologies that have enabled this growth, and the factors that are currently limiting further growth. These factors help to explain, for example, why Patagonian toothfish first hit the world market in the 1980s, as the technology for deep-sea fishing far from home improved, and stocks closer to home declined.

Chapter 3 looks at the structure of the contemporary global fishing industry. As an industry it is less concentrated than most other resource industries, and more beholden to governments both for subsidies and for long-term management. The current overexploitation of fisheries is not a problem of a few large companies pushing the market too far. It is a problem of too many participants, with too much capacity, chasing too few fish. Chapter 3 thus also looks at the market structure of both fishing and fish consumption, and at patterns of government support.

Chapter 4 discusses responses to overexploitation of the world's fisheries. It considers attempts by governments to overcome the issue structure and create rules that can at minimum prevent endangering specific species through overfishing, and at maximum contribute to the sustainable management of the world's fish resources. It examines both the failures of fisheries regulation, and its successes, including promising new regulatory approaches.

Chapter 5 changes focus, and looks at aquaculture, or fish farming. Whereas most of this book discusses capture fisheries, where wild fish are caught, aquaculture is an alternative source of fish. It has grown enormously over the past few decades, and now accounts for more than one-third of total global fish production. It is often portrayed as the solution to overfishing – if we can farm fish on a big enough scale, we do not need to catch them in the wild. And some forms of fish farming do hold promise as a sustainable source of fish. But not all. Some forms actually increase the pressure on capture fisheries. Chapter 5 therefore looks at the various kinds of aquaculture, their benefits and dangers, and their place in the bigger picture of the management of the world's fish resources.

Finally, Chapter 6 examines the question of what individuals, as both consumers of fish and as political agents, can do to promote sustainable fisheries practices. Although fish as a resource face numerous threats, consumers have demonstrated increasing ability, through their individual behavior and collective action, to influence conservation measures.

An examination of the geopolitics of fish as a resource demonstrates both the difficulty and the importance of the sustainable use of fish. The reasons it has been so difficult to sustain the world's supply of fish and avoid the ecological harm and human suffering that comes from poor management are clear. The world must use the knowledge gained from more than a century of trying to prevent fishery collapse to improve our ability to manage this resource, and others, in a socially and ecologically responsible way.

CHAPTER TWO

Growth of the Global Fishing Industry

Fishing and related industries are an important but relatively small part of the global economy, accounting for roughly one quarter of one percent of total economic output. They are the primary livelihood for tens of millions of people, and contribute significantly to the nutritional needs of the world's population. Who catches fish, what fish they catch and where, and how they do it, have all changed over the last century, and along with these practices, so has the health of the global fish stock.

Global capture fisheries reached a peak of about 90 million tonnes per year in the mid-1990s, and have declined slightly since then. Those catches result in just under 11 kilograms (about 24 pounds) per person per year globally. (If aquaculture is included those per capita figures increase by about half). Fish are thus an important, but still relatively minor, component of the global diet – the average person gets six percent of his or her total protein, and 15 percent of animal protein, from fish.[1] This percentage varies widely from place to place, however. In island nations such as the Maldives as much as half of people's total protein intake comes from fish, and nearly half the world's population relies on fish for at least 15 percent of animal protein consumption. In some inland countries such as Rwanda, Bolivia, and Bhutan, however, the figure can be well under one percent.

Over the past century, commercial fishing has become more industrialized and specialized, which has allowed for efficiencies and economies of scale. Vessels themselves often specialize, with some ships catching fish, others processing, and still others transporting. The size of the largest commercial fishing vessels has also, until recently, increased, as has the use of specialized technology for finding and capturing fish. Although the amount of fish caught has been on an upward trend for at least the past century, the fastest growth took place in the middle of the twentieth century, with catches doubling between 1950 and the mid-1960s, and doubling again between then and the mid-1980s. Since peaking in the mid-1990s, the total marine catch has begun a slight decline, as new improvements in fishing technology are offset by declines in fish stocks generated by the earlier rapid growth of the industry. This decline in marine catches has been, at least partly, offset by increases in freshwater catches, so consumption has been able to continue increasing. But the decline in the availability of ocean fish signals the overall unsustainability of global fishing efforts.

The growth in consumption of fish has increased slowly but steadily over the past half century. Until the mid-1980s, the growth in capture fisheries was faster than the growth in global population. Since then, capture fisheries have barely grown, but the rapid growth of aquaculture in the past two decades (see Chapter 5) has made up the difference. During this more recent period, most of the growth per capita in fish consumption has been in China, with consumption in the rest of the world, on average, changing little.

Even before the recent decline in global marine fisheries, however (a decline evident because more effort is required to catch the same amount of fish), global industrial fishing had been characterized by a boom-and-bust cycle of development. Fishing states expand to new regions, with fancier technology, and fish intensively until stocks of the target species in that region are depleted. They then move to a new species in the region, or to a new region with underexploited fish stocks, and the pattern repeats. Although technological development has allowed states to continue to catch fish as the stocks are increasingly depleted, there are no longer many pockets of underexploited fishery resources.

This chapter first examines the early history of fishing, and then discusses current fishing practices: Who fishes? How much? For what species? Where? It also explores the changing technology with which fishing has been done over time, including an examination of the current types of nets and lines used to catch fish and the effects of these gear choices and other fishing technology used both to catch more fish and, more recently, to help conserve fishery resources.

An Early History of Fishing

People have been fishing since prehistory. Fishing probably started in shallow waters in streams, lakes, tidal pools, and coral reefs, and with the simplest of technologies: rocks, spears, and even bare hands. Over time but still well before the beginning of written records, more efficient fishing technologies were developed, such as fishing lines and hooks, nets, and boats. As these technologies improved, people were able to fish for a greater variety of species, and were able to do so farther away from shore. In contemporary terms, the total global catch in this historical era was miniscule (although any attempt to come up with a specific number would be pure guesswork), and the overall effect on marine ecology negligible. It is, however, likely that as these early technologies improved some communities were able to overexploit the local stocks they depended on, and were faced with food crises.

Pre-modern technology in some cases developed to the point where impressively large creatures could be caught at sea. Native communities on the Pacific coast of North America hunted whales from canoes. In Sicily in the Roman era huge nets were used to catch bluefin tuna, which can weigh more than half a ton. For the most part, however, a large majority of fishing happened either inland (in freshwater lakes and rivers) or within sight of shore, until perhaps seven or eight hundred years ago, with the development in Europe of sail-powered ships designed for the open ocean. This new technology opened up vast new fishing possibilities.

In particular, technology allowed much greater access to two kinds of fish that had major impacts on European, and later global, history. One is small pelagic fish, or fish that live in the middle of the water column, such as herring. With the improvement of their seagoing technology, the Dutch greatly expanded their herring fleet in the fifteenth and sixteenth centuries. This fleet was well suited to long-distance bulk shipping, and became the basis for the central Dutch role in the large expansion of seaborne trade in the seventeenth century that marked the first real era of globalization. The other is demersal fish, or bottom-feeders, fish that live on the ocean floor, such as flounder or cod. The great cod fisheries off the North American Atlantic coast, from Cape Cod in Massachusetts to the Grand Banks off Newfoundland, were a fundamental element of the Atlantic slave trade in the seventeenth and eighteenth centuries. North American fishers sent cod to the sugar plantations of the West Indies, because it was the cheapest form of protein available to use for feeding slaves. The West Indies sent sugar to Europe, and Europe sent manufactured goods to North America. Without the cod, less sugar could have been produced in the Carribbean, fewer Africans would have been enslaved to grow it, and the economic development of white settlement in North America would have been significantly slower.[2]

Fishing continues in the places it has always been done, including in lakes and rivers, and close to shore. But these activities no longer account for the bulk of fish caught globally – freshwater fishing, for example, accounts for only one-tenth of the global total. The story of the expansion of modern fishing can mostly be told in two ways. The first is the expansion of the range of fishing, the increase of the distance from the shore and the depth of the sea in which people can successfully fish. The second is the expansion of fishing technologies, and the consequent ability of industrialized fishing to denude large areas of the sea of fish in increasingly short time spans.

Who Fishes the Oceans?

One way to tell the story of geographical expansion historically is to look at which are the primary fishing countries. Although people from coastal areas have always engaged in ocean fishing, the locus of large commercial fishing has changed over time. This geographic distribution of fishing power has changed from dominance by European states in the early centuries of ocean fishing, to current dominance by states in Asia and the Americas.

The states most central to the early expansion of industrialized fishing (in the fifteenth through nineteenth centuries) were Atlantic European states, with Mediterranean states increasing in fishing prominence during this period. Initially they fished in waters near their own coasts, but increasingly began traveling further away in search of other fish stocks. The most important area for fisheries expansion in this period was the North Sea, focusing on a number of species including herring. By fairly early in this period, the entire North Sea was being fished, including the areas hundreds of miles from shore. By the nineteenth century, fishing in the region was becoming intense enough that governments began to take notice of declines in fish stocks and in catches. Estimates between 1903 and 1907 suggested that for some species fishers were catching 70 percent of catchable fish in a given year.[3] The extent of overfishing in the North Sea became clear at the end of the First World War, in 1918. Most fishing activity there had ceased, as the war had made fishing too dangerous. Fishers returning to their trade immediately after the war were amazed by the recovery in fish populations created by the four-year break in fishing. This respite was short-lived, however, as newly increased fishing quickly depleted stocks, and by the early 1930s it was nearly impossible to catch fish commercially in much of the North Sea because of the over-harvesting.

Distance fishing was already practiced at this time. Perhaps even before the voyage of Columbus, Basque fishers from what is now northern Spain began fishing on the North American continental shelf, and Vikings from northern Europe fished off of what is now eastern Canada. European fishing off of the North American coast was one of the early drivers of European colonization of the area. After independence the United States began by fishing the abundant stocks of cod near its shores and quickly rose to prominence as an important fishing nation. Early US fishing focused on the demersal (ground) fisheries off New England, but the US fleet rapidly expanded and began to seek other fishing opportunities. The European and American dominance of global fishing continued through much of the nineteenth century, and into the twentieth.

In the early twentieth century Asian states became important fishing states. Japan's use of motorized trawlers in the 1930s (which intensified after the Second World War) allowed it to expand its catches and region of focus. China's growth in fishery production was increasing dramatically by the 1970s, and experienced its greatest increase in the 1990s. Although there is cause to doubt official Chinese statistics about catches during this period,[4] it is clear that both China's fishing technology and catches were indeed expanding quickly. Commercial fishing in the Philippines, Indonesia, and Thailand also expanded in the last quarter of the twentieth century, along with increases in fish processing in this region. Asian countries now account for more than half of the world's fish catches, and a considerably larger percentage of aquaculture production.

The Soviet Union also rose to fishing prominence in the second half of the twentieth century. In the 1950s, it became a Soviet national goal to develop the world's largest fishing fleet, a feat it accomplished, by some measures, in the 1980s. After the political collapse and subsequent economic transition of the Soviet Union in the early 1990s, Russia and the other successor states initially had few resources for global fishing efforts. Although Russian fishing has increased in the last decade, and Russia is currently in the top ten fishing countries, catches remain well below what they were in the Soviet fishing heyday.

Fishing by Latin American states grew dramatically beginning in the 1960s. Peru, Chile, and Ecuador have grown to fishing prominence, fishing primarily in the productive waters near their own coasts. The rise of fishing by Latin American states in their Pacific Ocean waters is important not just because of the volume of fish caught, but also because these states led the push for changes in how fishing in the high seas is governed. It was these states, as discussed below, that led the charge for exclusive economic zones, or EEZs, in order to protect fish stocks off their coasts from overfishing by long-range fleets from elsewhere, such as the growing fleets from the United States, Soviet Union, and East Asia.

The global pattern of fishing is currently fairly concentrated: twenty-three countries account for 80 percent of the world fish catch. For the last decade, China, Peru, and the United States have been the three largest capture fishing countries. By far the biggest fishing nation in the world is China, which accounts for catches of almost 15 million tonnes of fish per year, one-sixth of the global total. The concern about the reliability of fishery statistics from China, however, has led the FAO to report most of it statistics both with and without China. Given China's important role in both fishing and fish consumption, the unreliability of Chinese data increases the uncertainty about overall fishing behavior and fishery resources.

Peru is in second place, accounting on average for over seven million tonnes per year. The total for Peru varies from year to year considerably more than for other major fishing nations, because most of the catch by weight is from one species, anchovies, and the size of the anchovy run depends on climate conditions. Indonesia, the United States, Japan, India, Chile, and Russia round out the top 8, each with a total annual catch ranging from three to five million tonnes per year. Developing countries (including the unreliable Chinese figures) now account for almost three-quarters of the world's capture fishery by weight, and developed countries barely over one-quarter.

Developing country fisheries production increased most dramatically along with the designation of EEZs, areas that extend 200 nautical miles (approximately 370 kilometers) from a country's coastline within which it can exercise exclusive jurisdiction over marine resources. The 1982 United Nations Convention on the Law of the Sea (UNCLOS) formalized the practice, and as countries declared these zones they frequently ejected distant-water fishing nations that had been fishing in these areas and worked to subsidize and develop their national fishing industries. EEZs brought approximately one-third of the ocean under state control, and the majority of the global commercial fish catch.[5]

While much of the fishing in fresh water and near-coast

fisheries is still done in traditional, or artisanal, ways, fishing farther from shore and in deeper water is for the most part heavily industrialized. This kind of fishing is fairly evenly distributed across much of the world, with the largest vessels in Europe, Latin America, North America, and the Caribbean. There are currently four million fishing vessels, nearly 40,000 of which are larger than 100 tonnes and clearly engaged in industrial fishing. Many smaller vessels, however, may also be engaged in industrial fishing.[6]

Registry of fishing vessels has also changed over time as countries began to offer "open registration" (also known as "flags of convenience") to non-nationals, who could register their vessels in that location instead of in their home countries. Open registry states earn the revenue from registration and taxes, while offering vessel owners the advantage of lower fees and laxer regulations than prevail in their home state. Because ships are only bound by the domestic and international rules their registry states have adopted, fishing vessel owners may choose to register their ships in states that do not participate in regional fisheries management organizations; they are not, then, legally required to follow any fishing limits set by these organizations. Some flags of convenience, such as Belize in the 1990s and early 2000s, intentionally remain outside of fishing agreements in an effort to lure registrations by fishing vessels.[7] This practice has also made data collection more complicated, since ships may be registered in countries they otherwise have no connection to, and it is not clear that statistics collected based on registry bear any relationship to who is catching or consuming fish. The practice of open registration has also made international regulation more difficult, because these flags of convenience have often existed precisely to circumvent international rules.

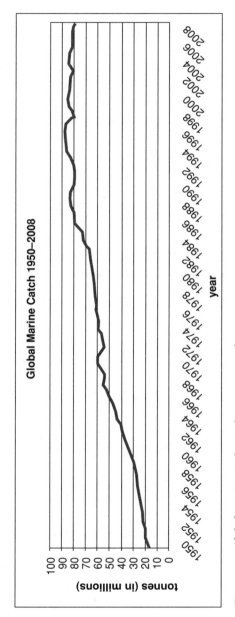

Figure 2.1 Global Marine Fish Catches 1950–2008

Source: FAO "Statistics – Global Capture Production," http://www.fao.org/fishery/statistics/en

How Much?

Fishing catches in the oceans have risen dramatically over the last century and a half. World catches in 1850 amounted, by very rough approximation, to about two million tonnes; by 1930 that amount had risen to about ten million tonnes, and had doubled again by 1950 to nearly 20 million. Catches increased steadily from that time onwards (the period during which aggregate statistics were regularly collected by the United Nations Food and Agriculture Organization), to a peak of nearly 90 million tonnes a year in the late 1990s, with current catches declining to closer to 80 million tonnes a year, as indicated in Figure 2.1.

For What?

The term "fish" here is used loosely, to refer to all forms of aquatic animals, including shrimps, mollusks, and even sea cucumbers. The biggest single category is small pelagic fishes (those that live in the open oceans, rather than near shore or inland, and in the middle of the water column, rather than on the surface or the sea floor), such as herrings, sardines, and anchovies. Those species account for almost one-quarter of the total world catch, followed by the cod family (including hake and haddock). Data on aggregate world catches are not, however, particularly precise – the categories of "miscellaneous" and "species not identified" account for over one-third of the total in global catch statistics, and even those statistics likely miss some catches.

Historically, the early target species during the European dominance of industrial fishing were cod and herring, with a dramatic expansion in the twentieth century of the types of fish caught and consumed. Tunas and billfishes – larger, faster, further-traveling top predator fish species – rose to prominence and now comprise nearly ten percent of global marine catches.

The species that we see most often on menus do not necessarily account for that large a proportion of the total global catch. All species of tuna together account for just over six million tonnes, shrimp for just over three million, and salmon for just over one million tonnes. In part this disparity can be explained by aquaculture. These figures are for capture fisheries only, and both shrimp and salmon are farmed in large volumes. But most of the disparity points to different market values for different kinds of fish. The figures given here are by weight, not price, and the species we hear about most tend to be the most expensive. The highest volume catches tend either to be consumed locally or to be processed into generic fish products.

The largest fishery in the world by volume is the Peruvian anchoveta. In the best fishing years for the species it accounts for ten percent of all global capture fishing. That fishery also experiences the greatest degree of volatility, because it is influenced by the El Niño Southern Oscillation (ENSO), a periodic climatic pattern in the Pacific that changes ocean temperatures and pressures and thereby affects fish populations. In years with a strong El Niño, stocks decrease dramatically. Other high-production fisheries include (in current order of importance) Alaskan Pollock, skipjack tuna, Atlantic herring, blue whiting, chub mackerel, Chilean jack mackerel, Japanese anchovy, largehead hairtail (a species of cutlassfish), and yellowfin tuna.

Although some commercially caught fish is consumed in the country that caught it, fish is a major trade commodity, particularly in developed countries. Overall, just under 40 percent of global fishery production (including both capture fisheries and aquaculture) is exported. Many developed countries both export and import large quantities of fish, some of which may in fact be re-exported. Currently the top exporters of fish (again including both capture fisheries and aquaculture) are China, Norway, Thailand, the United States, Denmark, Canada, Spain, Chile, the Netherlands, and Vietnam. Japan and the United States vie for the top spot in the list of importers, with Spain, France, and Italy rounding out the top five. Note that these figures are based on the port where fish are landed, not on whose waters the fish was caught in. So if a fish is caught in Senegalese waters by a Spanish vessel and ultimately eaten in Japan, it will probably be counted as a Spanish export, not a Senegalese one.

Where?

Most of world's fish catches come from the oceans. Pelagic fishing – fish caught on the open ocean rather than near the shore – became the primary source of world landings of fish beginning in the 1950s, although distant-water fishing had been practiced for centuries. Nonetheless, 90 percent of ocean fish are caught in EEZs, within 200 miles of the coast. The extent of near-shore versus open ocean fishing varies by region.

Inland (freshwater) fisheries, although accounting for only roughly one-tenth of fish catches worldwide, are currently growing faster than marine catches. Given that most of these fisheries were already well-developed and that freshwater fish are under increasing strain from pollution and habitat loss, this growth suggests that inland fisheries are facing a serious potential problem of overexploitation. Most of these catches are consumed domestically, and accurate catch and stock estimates are difficult to obtain (especially since approximately 95 percent of inland capture fishing is done by developing states, and most of that by China,[8] whose statistics have been shown to be questionable), so this aspect of the industry is not considered in depth in this analysis.

Whereas the North Sea was the location of the most fishing activity in the early centuries of commercial fishing, currently the Pacific Ocean is the location of the greatest capture fisheries production, accounting for about three-fifths of fish captured at sea (i.e. not including inland, or freshwater, fish). The Northeast Atlantic (including the North Sea) is the next biggest source of marine catches, although catches there have been gradually declining over the course of the past decade. Catches in the Indian Ocean have recently been increasing, although from a much smaller base, accounting currently for about 15 percent of the total. The progression from one fishing area to another often represents the increasing intensity of exploration of new fishing areas as others are increasingly fished-out. Few untapped fishing resources exist.

How?

Historically, fishing was a low-tech operation. Fishing can be done in hand- or wind-powered vessels, with a small net or a hand-held line. Even now, most people engaged in fishing or fish farming do so in low-technology, small-scale, artisanal ways, along coastlines and in inland waterways. Even among motorized fishing vessels the vast majority are smaller than 12 meters in length. But although these local fishing operations employ many people, they have a much smaller impact on the state of the actual resource than do the increasingly large and technologically sophisticated industrial fishing operations. Currently there are 2.1 million motored fishing vessels, but only a small percentage of those fish on an industrial scale.[9]

Mechanized fishing began at the end of the nineteenth century, as fishers from Britain and Europe left depleted local fishing grounds to venture further away, in larger, steam-powered vessels. One of the major causes of transformation in global fisheries over the last century or more is changing technology. Innovations in shipbuilding allowed the creation of large vessels that could stay out at sea for months at a time or even longer and seek fish further from shore. By the sixteenth century, and possibly even earlier, ships travelled across oceans to catch fish. The fish would be salted shortly after being caught, preserving them potentially for years. Salt cod is still a staple food in places like Portugal and Spain that traditionally fished far from local waters.

Technology made transport and fish catching more efficient. Technological advances in gear allowed for larger nets or longer lines to be used. Processing facilities on board allowed catches to be prepared for storage and frozen so that the ship would not have to return to port to offload its catches, and could therefore travel further in search of valuable species. Simultaneously, developments on land made transport faster and less expensive, which made it easier for fish to be sold and consumed far from where they were caught. The same refrigeration that allowed for storage on fishing vessels also allowed catches to be shipped further inland, increasing the consumption of ocean fish among people who did not live near an ocean, and the increasing availability of air transport meant that fresh fish could be shipped quickly from where it was caught to where it would be consumed. Technology developed in military contexts, such as sonar, radar, and global positioning systems (GPS), came to be used in peacetime to find fish more efficiently. The dramatic technological advancements, especially in the second half of the twentieth century, changed the face of ocean fishing.

Fishing Vessels

The trend over time has been for fishing vessels to increase in size and specialization. The move from sailing to motorized vessels for fish catching increased possible size and also speed. The development of factory trawlers, vessels that can catch and process fish on board and thus stay out to sea for long periods of time, followed from whaling ships that processed their catches at sea.

A related development is fish carriers, vessels intended to take fresh or frozen fish from the vessels that catch or process it to the markets or processing areas; with rising fuel costs worldwide these carriers are an economic way to transport fish so that the fishing vessels themselves can remain out in distant fishing areas for a longer period of time without having to travel back to a home port. Currently there are approximately 740 such vessels worldwide and they can be enormous. Russia and China have the greatest capacity in fish carriers currently; Belize, Cyprus, and Panama also flag especially large fish carriers, although these ships are likely to be owned by people from other countries.

The number of large commercial ships fishing on the ocean began increasing dramatically in the 1970s and 1980s, as technology and subsidies (discussed further in Chapter 3) increased the ability of fishers to make use of larger ships. Vessel size (among the major commercial fishing vessels) also increased steadily over time, reaching a peak in 1992, but decreasing dramatically since then, as states began to limit capacity of fishing vessels out of concern for fishery conservation and in response to the cost of operating the biggest vessels. Since then, the numbers of vessels have held steady or even increased slightly, but they are, on average, smaller. Japan and the European Union in particular decreased both size and numbers of fishing vessels gradually in the 1990s. Large (over 100) gross tonnes fishing vessels registered in Japan decreased in number from 2,762 in 1992 to 1,562 a decade later, and the EU fishing fleet during this area decreased on average between two and three percent a year.

Of the 4 million fishing vessels on the oceans, nearly 2 million fishing vessels are small and un-motorized, used by artisanal fishers. The other two million vary in size dramatically from 10 meters or smaller to the largest fishing vessel currently in existence, at 144 meters in length (and 22,055 gross tonnes). Beyond these numbers, there are numerous boats used in recreational fishing that are rarely included in official estimates.

Fishing Gear

Early fishing efforts involved simple technologies – often a line cast from an unmotored boat – but gear types have grown more sophisticated over time and able to catch larger numbers of fish with greater precision. The growth in fishing nets is one such endeavor. Artisanal fishing often makes use of hand nets with which fish can be scooped directly from the water, or cast nets, small nets that are thrown short distances by hand and then hauled back in.

Gillnetting, used initially in artisanal fishing but expanded by the mid-1800s into international commercial fishing, involves hanging a large net vertically from the surface. Fish try to swim through, and catch their gills in the net, and are then unable to move backwards or forwards. Historically gillnetting was a primary method for catching herring in the North Sea, and cod have been caught this way as well. Salmon are currently frequently caught by gillnetting, and fish species from throughout the water column can be targeted this way, because gillnets can reach from the surface to the sea floor.

Driftnets are essentially unanchored gillnets and, as such, can be used to target any species that can be caught with a gillnet, although they were often used to catch high-value species such as tuna and swordfish, as well as squid. The nets are released and allowed to move about in the ocean until they are hauled in with their catches. These nets can be up to several miles long. Bycatch is a particular problem for driftnets, because these nets ensnare any fish that come into their path. Because of the indiscriminate catch from driftnets, environmental groups organized campaigns against their use. The United Nations in 1992 adopted a resolution opposing their use on the high seas, and the United States imposed economic sanctions on countries that refused to adopt this ban in international waters. Driftnets are still sometimes used, although their size is generally regulated, within nationally controlled waters.

A related net type is a purse-seine net. These are used to encircle a school of fish; the net is then pulled closed and the catch hauled in. The primary method of catching yellowfin tuna in the Pacific Ocean was, until the late 1970s, by encircling dolphins with purse-seine nets. Yellowfin tuna frequently school with dolphins, which – unlike the tuna – can be easily seen at the surface. So a reliable way of finding schools of yellowfin was simply to find and encircle dolphins. Despite efforts to "back down" the nets and allow dolphins to escape, dolphin mortality skyrocketed, and US, and then international, rules regulated, and greatly decreased the use of this approach.[10] Purse-seines are used to catch schooling fish such as (in addition to yellowfin tuna) herring, anchovies, and salmon. There are other variations of seine nets, but the commonality is that these nets float on top and hang vertically with attached weights.

The other kind of fishing net is a trawl. This kind of net is drawn behind a moving vessel and gathers up the fish it ensnares as the vessel moves through the water. (Trawling can also be done by drawing the net behind two vessels.) Trawling is done in different parts of the ocean. Midwater trawling catches pelagic fish such as anchovies or tuna. Bottom trawling involves dragging the net across the sea floor, in search of groundfish species like halibut or flounder; in addition, some pelagic fish, such as blue whiting or some species of rockfish, can be caught through bottom-trawling. Although all trawling leads to bycatch, bottom-trawling is particularly harmful to the ecosystems in which it takes place, because it can harm structures on the ocean floor, such a deep water coral reefs and seaweed, on which fish and other aquatic creatures depend. It also stirs up sediment that clouds the water, and can stir up pollutants, causing harm to the species that live there. Trawling has been done for centuries, but, as with other netting techniques, has been made more efficient over the years.

Technological developments have made net fishing much more lethal to fish. One of the major developments in both net and lines was innovations in type of filaments. Nets made from synthetic fibers last longer and require less maintenance, and new microfilament nets are much less visible to fish and therefore result in bigger catches. The other impact of new fibers in nets is that nets that are not retrieved often move around the ocean for years as "ghost nets," damaging ecosystems and killing fish that then cannot be used for human consumption or cannot reproduce to help replenish fish stocks.

Innovations in the first half of the twentieth century added mechanization to the setting and withdrawing of nets, making the process more labor efficient and able to haul in larger catches. Improvements in motors for net winches continue to help the efficiency of net fishing.

The other major type of fishing is done with lines. As with net fishing, line fishing ranges from a simple handheld fishing line (perhaps attached to a pole) to industrial processes that bait thousands of hooks mechanically and reel miles of line out and in. Traditional deep-sea fishing involved one hook per line, and one line per person, much as is the case with sport fishing today. Large ocean-going fishing vessels would have dozens of fishers lining the sides of the boat, holding lines. Contemporary industrial line fishing has increased the number of hooks per line, and decreased the number of people required to operate the lines.

Longlines are, as the name would suggest, long fishing lines to which baited hooks are attached from subsidiary lines that hang down from the main line. These lines can range from only a couple dozen hooks to many thousands over a line that extends for several – up to fifty – miles. The hooks can be baited and reeled out by hand, but technical advancements in the last few decades have made it easier to automatically bait and to lay and reel in longlines, increasing the efficiency of fishing.

As with trawl nets, longlines can be placed at different locations in the ocean, depending on where the target species live. Pelagic longlines are set at the sea surface and used to catch fish swimming fairly near the surface, such as tuna or swordfish. Demersal longlines operate toward the sea floor, to catch groundfish like cod, halibut, or toothfish. They can either be set at the surface, with weights to sink the lines to their target depth, or released underwater from the body of the vessel.

Longlines catch non-target species, including sea turtles and seabirds, as well as juvenile fish of the target species. Some modifications, such as circle hooks, in which the end of the hook curves inward toward its straight section, dramatically reduce bycatch of sea turtles and sea birds. Demersal longlines that are released underwater are less likely to result in bird bycatch.

Finally, some smaller-scale fishing approaches can also have dramatic effects on the ecosystem and thereby long-term effects on the availability of fish. Blast fishing is a practice in which dynamite or other explosives are ignited, killing or stunning fish in an area and making them easy to harvest.

This practice has existed for nearly a century, and is primarily practiced near coral reefs in southeast Asia and off the coast of East Africa. It has been outlawed in some areas, because of the devastation it causes to the surrounding ecosystem or habitat (especially since blast fishing is frequently practiced near coral reefs, which play important roles as fish nurseries). It is nevertheless hard to police, in part because it is done by many small-scale fishers rather than a few big ones, albeit with large impacts on marine ecosystems. A 1999 study suggested that, despite its illegality at the time, 12 percent of Philippine fishers engaged in blast fishing.[11]

A similar problem is poison fishing, the practice of releasing cyanide (or bleach) into a reef area to stun fish for easy capture. Fish harvested in this manner are often taken live for the aquarium trade or as fresh seafood (sometimes shipped live). Grouper and Napoleon wrasse are frequently caught in this manner, especially in southeast Asia. In addition to whatever effects the poison may have on the fish themselves, this practice almost certainly harms reefs and other species in the area where it is used. Although this process is low-tech and not used in industrial fishing, it can nevertheless have an impact on availability of fish on a broader scale. A similar fishing process is used in parts of Africa, such as fishing in Lake Victoria in Tanzania, where various poisons (such as pesticides) are used to kill fish, which then float to the surface and can be easily harvested. Although this process does less ecosystem damage than blast fishing, it can have serious health effects, because the fish continue to carry the poison that killed them.

Fish-Finding
Fish-finding technology owes its legacy to wartime developments. Sonar was developed to locate submarines during the First World War, and was in regular military use by the

Second World War. It is one of several types of acoustic technologies that began to be applied to fishing in the post-war period. Schools of fish can be detected via sound waves emitted into the water, since fish have a different density than water. Sonar can also be used once nets have been deployed (sometime via sensors or emitters on the nets themselves), to determine the optimal location of the net. In addition, fishers use sonar and related sound technologies to determine characteristics of the seabed, and water depth, in ways that can guide their decisions about fishing location.

Spotter airplanes also help with fish-finding in modern factory-fishing contexts. Pilots locate fish schools as seen from the air and direct associated fishing vessels to the most productive schools. They can also help vessel crews most effectively determine where to set nets. They have played a key role in some fisheries in recent times. For instance, from the mid-1980s to mid-1990s more than 85 percent of net sets in the Atlantic Menhaden fishery were guided by spotter airplanes. Use of spotter airplanes has been determined to increase the efficiency of catch in a fishery by at least one-third. In some fisheries, such as the Atlantic bluefin tuna fishery managed by the International Commission for the Conservation of Atlantic Tunas, the use of spotter airplanes has been banned.

The global positioning system (GPS) is a directional tracking system, originally developed by the US Department of Defense, that makes use of satellites to allow users of GPS receivers to determine their exact location on the globe. This technology was originally used only for military purposes, but was made available by the US government for civilian use beginning in the 1980s. By allowing fishing vessels to determine their precise location at any point in time, this technology made it much easier for vessels to find and return to productive fishing areas. Vessels equipped with GPS receivers have been shown to be measurably more productive at catching fish.

Other Impacts of Technology
TRANSPORTATION AND PRESERVATION
Some of the most important impacts of technological development on global fish stocks have little to do with how fish are caught but rather with how they are transported after they have been caught. The ability of ships to process catches at sea allowed for preservation (initially drying, but later refrigeration) that meant that the markets for fish could be far from where they were caught.

The increasing availability of refrigeration also made other long-distance transport, inland in trucks or trains, possible, and ocean fish became a staple food for people far from coastlines, thereby increasing the number of possible consumers. The increasing speed and decreasing cost of transportation also meant that fresh fish could be shipped further from the location in which they were caught or landed. Fast transport of sushi-grade fish to Japan, beginning with the use of jet airplanes for such purposes in the 1970s, put enormous pressure on the Atlantic bluefin tuna fishery, as the price fishers were paid for a single bluefin tuna skyrocketed. The recent record for bluefin tuna was a fish that sold at the Tsukiji market in Japan for $173,600 or $391 a pound.[12] When a fish can fetch that kind of price when fresh, the air transport to ship it becomes cost effective.

RESEARCH AND CONSERVATION
Technology is also helping those who manage fisheries. Computers assist in modeling population dynamics, at a level of sophistication that would have been impossible without them. Once these models exist, the effects of different ranges

of fishing behavior can be tested before regulatory decisions are made. Sonar seabed mapping technologies can record characteristics of seabed habitats to predict distribution of fish in an area. Sonar can also be used as a tool for conducting assessments of fish stock abundance. Likewise, airplanes can be used to spot illegal fishing instead of assisting vessels in locating schools of fish.

Bycatch reduction devices are technological innovations used to reduce the amount of bycatch from the types of large-scale fishing operations that most produce it. Some operate physically, by refusing non-target species entry into a fishing net or guiding them out. Turtle Excluder Devices (TEDs) on shrimp fishing nets, for instance, allow sea turtles, which have a tendency to get caught in shrimp trawl nets, to swim out of an escape hatch in the net.

Because bycatch is by definition the catching of non-target species, the use of bycatch reduction devices should make fishing operations less wasteful. Their mandated use is sometimes resisted, however, for fear that any mechanism to either release or deter bycatch may also have that effect on target species. Shrimp can be lost through the hatch that allows turtles to escape, and for some fish species the use of bycatch-reducing circle hooks may reduce the efficiency of target species catch, in some cases considerably (although some studies suggest that for tuna and swordfish they may not).

Vessel monitoring systems (VMS) are also becoming more common in fishing regulation. A satellite (or sometimes radio) tracking system is mounted on a fishing vessel and relays real time information about its position and speed to either regulatory agencies within registry states or to regional fisheries management organization headquarters. This technology is used in domestic fishing regulation such as scallop fishing in the United States and anchovy catches in Peru; it is also required internationally by some regional fishery management organizations, such as in the Patagonian toothfish/Chilean sea bass fishery in Antarctic waters and for Southern bluefin tuna caught under the auspices of the Convention for the Conservation of Southern Bluefin Tuna. The information provided by this technology can be used to certify where fish were caught and thereby determine whether fishing happened within the proper regulatory area. Despite instances of falsifying tracking data or moving catches to unregulated areas, it seems likely that this tracking technology has a deterrent effect that keeps vessels fishing in the areas where they are allowed to be. Without this kind of tracking or monitoring technology it can be quite difficult to keep track of what individual vessels are doing on the vast ocean.

Conclusion

Despite its reasonably small contribution to the global economy, fishing is of economic importance to the people who engage in it, and can provide an important source of protein for the populations of coastal or island states. Fishing practices have changed over time. Dominance in global capture fisheries by European states yielded to dominance by those in the Americas and now Asia. Fishing gear and fishing vessels have grown more specialized and more technologically sophisticated over the last century, allowing people to find and capture fish much more efficiently. And transport and preservation technology has allowed fish consumption to expand globally, far from the locations in which the fish were originally caught. After increasing for decades (and probably centuries) world catches have now plateaued and may have even begun to decline, because of this increased fishing effort and consumption.

Gear that helps fishers capture more fish with less effort has effects beyond just leading to overexploitation of target species. Huge nets that can scrape across the sea floor or be towed through water columns capture fish indiscriminately and unwanted or disallowed bycatch may simply be discarded. Fishing lines miles long with thousands of hooks can catch and kill seabirds and other endangered species.

But technology that helps deplete fisheries can also be used to help conserve them. Sonar and other tracking and mapping technologies can be used to plot out the best areas to create marine reserves or conservation zones, or to track tagged fish for scientific study. Satellites can ensure that fishing vessels are following national or international rules. The increased ability of people to catch fish is a major problem for global fisheries, even more so in the context of incentive and regulatory structures, discussed in subsequent chapters, that encourage them to continue to do so even when global fish stocks are declining.

CHAPTER THREE

Structure of the Fishing Industry

Only part of the story of global fishing can be told by understanding which countries' nationals fish which areas of the oceans with what technology for which species. The structure of the fishing industry helps explain why people or businesses fish and why they do it in the way or the places they do. Political, economic, and social factors intertwine to influence the extent and manner of resource use globally.

The geopolitics of most resources is characterized, at least at some point in their supply chains, by industrial concentration. The oil industry is dominated by a few huge multinational corporations that play key roles in global distribution, and one government that has hegemonic power in supply. The supply of many key minerals is dominated by a handful of producers. Even in agriculture, an industry in which there are billions of primary producers, the supply of seed to much of the world is dominated by a few corporations with global reach. Global fisheries are different. They are similar to agriculture in that primary production is broadly dispersed. No individual fishing company catches a big enough share of the world's fish to have real market power. While the global distribution system of fish products is becoming slightly more concentrated, the industry as a whole remains much less so than is the case with most other primary resources.

But at the same time, the industry looms larger on the agenda of many governments than its economic size would seem to warrant. In countries with established fishing industries, fishers often wield political influence disproportionate to the economic importance of fishing. In countries with untapped stocks, the creation and expansion of the fishing industry often seems like a costless means of economic development. This potential, and the eventual politicization of the industry, often leads governments to subsidize fishing to a degree that creates significant overcapitalization, in the form of more fishing vessels, and more technology, than fish stocks can support. This overcapitalization is one of the key causes of overfishing and of depletion of the world's fisheries.

This chapter addresses these features of the structure of the global fishing industry. It begins by noting some basic statistics about the industry, and drawing a distinction among industrial, artisanal, and recreational fishing, its three primary elements. It then focuses on the role of politics in supporting the industry (as opposed to the role of government in regulating the industry, which is covered in Chapter 4), and the ways in which parts of the industry move internationally to seek out the most favorable government treatment. The final section looks at the question of where fish go after they are caught. It distinguishes between the high end of the market, where fish are valuable enough to be flown around the world to be eaten fresh, and the low end, where low cost matters more than species.

Industrial, Artisanal, and Recreational Fishing

The fishing industry, broadly defined, is a rapidly growing source of employment and income. More than 43 million people are employed in primary production in capture fisheries and aquaculture, with an additional 4 million employed occasionally. Aquaculture, discussed further in Chapter 5, is the fasted growing sector of fisheries employment. Capture fishing and aquaculture, however, account for only a small fraction of the economic and employment reach of the fishing industry broadly defined. Fish processing, marketing, and other service industries are central to global fishing worldwide. The United Nations Food and Agriculture Organization (FAO) estimates that for every fisher there are approximately four others in these secondary operations, for a total of 170 million people employed worldwide in the fishing industry.[1] Nearly eight percent of the world's population is dependent for its livelihood on this industry, supporting additional members of their families and households.

One of the most notable features of the global fishing industry is the wide range of scales on which it is practiced. At one end of the spectrum is a traditional fishing village in a poor country, where fishers use hand-made nets and small hand-made boats, or no boats at all. These are artisanal fishers, who have little capital, can fish only close to home, and fish primarily for local consumption. At the other end of the spectrum are deep-sea trawlers and longliners, hundreds of feet long and costing tens or, in some cases, hundreds of millions of dollars. This industry is highly capital-intensive, with global range and access to global markets. Off this spectrum entirely is the recreational fisher, who in some cases will spend hundreds or thousands of dollars on travel and lodging for the chance of catching a few fish that may even be released after being caught. These three kinds of fishing follow different economic logics, but are interrelated in their effects both on the marine ecosystem, and on the political economy of fishing.

Most of this volume addresses industrial fishing, because it has the greatest impact on the global supply of fish. But it is important to pay attention to artisanal fishing as well, because it is the most widely practiced form of fishing globally. The United Nations Food and Agriculture Organization (FAO) characterizes it as fishing conducted by households rather than commercial entities, using minimal capital and energy, and small vessels that make short trips close to shore. The fish caught in this manner is used mostly for domestic consumption, although the FAO notes that local definitions vary considerably, and can include fish sold commercially and even exported.[2] Although much more fish by volume is caught by industrial fishing processes, artisanal fishing employs far more people.

Recreational fishing differs from artisanal and industrial fishing in that economically it is a form of tourism rather than a form of resource extraction. As such, a given number of fish can generate much more economic activity than commercial, let alone subsistence, fishing. Much recreational fishing has fairly little economic impact – fishing off the local dock does not do much to support the local community. But fishing for trophy species, such as marlin in the Atlantic or salmon in the Pacific Northwest, can require long-distance travel and lodging as well as boat rentals and permits. In particular instances fishing for sport can become big business. It has been estimated that the recreational salmon fishery in British Columbia contributes more to the province's economy than the commercial salmon fishery,[3] surprising though that statistic may seem given the political salience of the commercial fishery in the province. The economic value of recreational fisheries can be undermined, however, by bad management of commercial fisheries or even by overfishing in subsistence fisheries. The effects of declines in recreational fishing opportunities may in some cases cause more overall harm to local economies than do decreased catches in commercial or artisanal fishing.

The Politics of Fisheries Support

Fishing attracts a level of government interest and involvement well out of proportion to its economic importance as an industry. For a few small countries, particularly islands with huge exclusive economic zones, such as Iceland or Vanuatu, fishing is critical to the national economy. For most countries it is of only marginal economic importance. The availability of figures for employment in fishing is erratic, but even for countries, such as Norway or New Zealand, in which fishing is a relatively important part of the economy, capture fishing employs less than one percent of the population. For large European countries with active fishing industries, such as Spain, Italy, and France, less than one-tenth of one percent of the population is employed in capture fishing.[4]

And yet fishing looms large in the national politics of many countries. Fishing unions are strong in many European countries. Fishing interests are often represented at the cabinet level in national governments, either with their own department or in conjunction with oceans or agriculture. Canada, for example, has a Department of Fisheries and Oceans, even though the value of fish landed is barely over $2 billion in an economy of more than $1.2 trillion (Canadian dollars in both cases). The European Union has a Directorate General of Maritime Affairs and Fisheries, the equivalent of a cabinet-level Department of the European Commission, the governmental bureaucracy for the world's second-largest economy. The exception among active fishing countries in the developed world is the United States, where the National Marine Fisheries Service (recently renamed the NOAA Fisheries Service) is a third-tier bureaucracy, part of the National Oceanic and Atmospheric Administration, itself part of the United States Department of Commerce. But much of the regulation of fishing in the United States happens at the state level, and fisheries bureaucracies are more prominent in states with major fisheries, such as Alaska, Massachusetts, and Washington.

Why the political importance of an industry that is relatively small, both in terms of monetary value and in terms of employment? Four factors help to explain this phenomenon. The first is the fact that any open-access resource needs to be regulated, or it will be over-used. An open access resource is one that anyone can gain access to without paying for. Fish in the wild are free (in the economic sense), and one need only catch them to add value. Governments therefore generally need a regulatory apparatus to limit the amount of fish caught, either through licensing or quotas. The fact that the resource is free for the taking, and needs only to be caught to be economically valuable, also explains why is it often used by governments to promote economic development, as discussed below. The open-access character of fisheries explains why governments need regulatory bureaucracies for them, such as the NOAA Fisheries Service. It does not, however, explain why these bureaucracies are so much more prominent in so many countries than the economic importance of the industry would seem to warrant. Explaining the political salience of the industry requires looking at the other three factors.

One is simply history. Fisheries historically were, for many countries, a more economically important industry than they are today. Particularly for large countries engaged in industrial fishing, the industry has not increased in output as quickly as the economy has more broadly, and employment in the fishing industry has often declined. In other words, when the bureaucratic structures governing fisheries were created, they may not have overrepresented the economic importance of fisheries. But those bureaucracies have remained in place even as the importance of the industry has declined.

Another factor is the frequent geographic concentration of fishers in economically disadvantaged areas. The fishing industry tends to be located in particular regions and ports. As fishing technology improves, fewer fishers are needed to catch the same amount of fish. Over time, then, employment in the industry tends to decline, and the geographic specificity of the industry means that there are rarely other local sources of employment able to absorb surplus fishers. This geographic concentration also means that fishing communities often have a concentrated political voice, out of proportion to the size of the industry. And finally, the small economic size of the industry means that it can often be bought off at a relatively modest cost. For example, the United States subsidizes its fishing industry by a little over a billion dollars a year. Against the scale of the industry this subsidy is a huge amount of money, equal to as much as one-third of the total value of fish landed by the industry in domestic ports. But against the scale of general government spending, or government assistance to other industries like agriculture (on an annual basis) or finance (when the industry needs to get bailed out every couple of decades or so), it hardly registers.

The final reason that fishing as an industry is more politically salient than its economic importance would suggest is the cultural and historical role of fisheries in both national and regional consciousnesses. Fishing evokes the romance of the sea, and the lore of the hunter. Fish and fishing have become central to the cultural self-image of some regions. Salmon, a staple catch of the native peoples of Pacific Canada, are central to the cultural imagery and history of British Columbia. Cod play a prominent role in the history of Massachusetts; a wooden model of a codfish is suspended in the central rotunda of the state legislature. Although in both of these locales a predominantly urban population has little interaction with fishing or the sea (aside from going to the beach), the fish that supported the economy in its earliest days remains a part of the local self-image.

In some traditional fishing areas these last two factors, economic decline and cultural embeddedness, combine to create a strong local or regional political voice for government support of the fishing industry. If support is not offered for maintaining the industry onsite, entire communities will disappear, with the loss both of economic infrastructure and of a rich local heritage based on fishing and the sea. Examples of this effect can be found in places like Galicia in Spain and Newfoundland in Canada. In both places local economies have been based on the fishing industry for centuries, with no sign of other sources for similar levels of employment (although in Newfoundland the discovery of offshore oil is beginning to change employment options). In both places the industry is sufficiently regionally concentrated that it can affect election outcomes, both regionally and, particularly in Parliamentary systems, nationally. As such, in both regions it makes political sense for the government to support the industry, even when it makes neither economic nor environmental sense.

Subsidies

These four factors drive the governments of many countries to involve themselves in the fishing industry in ways that undermine efforts to manage fisheries sustainably. Most government involvement takes the form of subsidies, broadly understood. Economic theory tells us that subsidizing an industry has the effect of keeping more people and capital in it than would otherwise be the case. In other words, subsidization leads to overcapacity. The fact that fish, from the perspective of many governments (and from the perspective of calculating figures like gross domestic product, or GDP) are available for free, and only need to be caught, also means that governments are prone to using them as a means to economic development. This development many be done by subsidizing existing fishers to increase the capital intensity of the local industry, or it may involve creating a new local industry that did not previously exist.

Government subsidy of fishing can, and does, come in many forms. Most types of subsidies have the effect of keeping more fishers employed than could be supported by an unsubsidized industry in an open market. The political demand for government support of the industry remains strong, because subsidizing fishers allows them to remain in the industry and in the same place; the political constituency for such support is thus kept intact. Other forms of subsidy have the effect of increasing the amount of capital in the industry beyond what the market would otherwise support. This excess capital can be used to build more or bigger fishing boats than would otherwise be built, or to increase the level of technology on existing or new ships beyond what would otherwise be purchased. This new technological sophistication includes many of the elements discussed in the previous chapter; electronics such as sonar and radar for spotting fish, Global Positioning Systems (GPS) for identifying precise positions at sea, as well as improved mechanical systems such as better engines, stabilizers, fishing gear, and on-board fish preservation and processing facilities. More boats lead straightforwardly to the ability to catch more fish. Boats with more advanced technology also lead to the ability to catch more fish.

The point of the improved technology is to make the vessels more efficient, and more efficient means the ability to catch more fish with the same size boat and crew. So a program to modernize a fishing fleet has the effect of increasing fishing capacity. Because subsidies often increase yield rather than (or in addition to) increasing the number or size of vessels *per se*, even efforts by governments to restrict the size of their fishing fleets often fail to have the desired effect if they are also subsidizing the modernization of the fleets. For example, in response to clear signs that its tuna fleet was too big and too old to be economically viable, the Japanese government created a program in the 1980s to both shrink the size of the fleet, by buying some ships from fishers and taking them out of commission, and subsidize the modernization of the rest. Improved technology (ranging from better engines to better radar) meant that each vessel could catch tuna considerably more efficiently. The improvement in yield generated by the modernization program offset the decrease in the size of the fleet from the buyback program. Moreover, many of the decommissioned vessels ended up in Taiwan or China, instead of being scrapped. So in the end considerable government money was invested in a program that neither reduced the pressure on tuna stocks nor improved the economic viability of the Japanese tuna fishing industry, and actually expanded the number of vessels fishing globally. Similarly, an EU subsidy program between 2000 and 2006 intended to decrease fishing capacity supported the scrapping of 6000 vessels while also contributing to the construction of 3000 new vessels and modernization of another 8000, leading to an overall increase in fishing capacity.[5]

Many of the forms of subsidization of the fishing industry are direct and obvious. Examples include grants to either build new or modernize existing vessels, or assistance with the operating costs of the fishery. For example, before the creation of EEZs, Senegal subsidized engines (by making them tax-free and offering easy financing) sold to small-scale fishers, so that they would be better able to compete with others – primarily European vessels – fishing off its coast.[6]

Assistance with operating costs can include below-cost or guaranteed access to insurance, compensation for damaged gear, deferral of income tax, or loans for operating costs. Subsidizing fuel is one of the most common types of operational subsidies; vessels that do not have to pay the full cost of (or the full tax on) fuel can afford to use more than they otherwise would. In addition to their effect on fishing behavior, fuel subsidies have other perverse environmental effects, by increasing greenhouse gas emissions, and thereby contributing to climate change. After its initial period of engine subsidization, Senegal turned to subsidized fuel for fishing vessels, in the hopes that it would keep fish prices low enough for the local population to consume. These fuel subsidies encouraged vessel owners to install even more powerful engines, and vessels then traveled further and stayed out for a longer period than they previously had. This subsidy is credited with increased catches that came from making use of new fishing areas and developing a purse-seine industry dependent on motorized vehicles. Most of this fishing focused on the export market, however, despite the intention of the fuel subsidy to keep prices low for domestic consumption.[7] In this instance, the modernization of the Senegalese fleet, made possible by subsidies, rapidly led to a depletion of the stock it targeted.

The provision by governments of infrastructure that is either necessary to the successful operation of a local fishing industry, or that makes fishing operate more smoothly, can also be seen as a form of subsidy. Common examples include the maintenance of port and navigational facilities, and of the regulatory structure of fishing. For example, Japan allocated US$ 5 billion annually between 1994 and 1999 for port expansion and maintenance specifically for the needs of the fisheries sector.[8] Commercial fishers rarely pay the full costs of running national fisheries departments or administrations, and rarely pay any of the costs of running international fisheries regulators (regional fisheries management organizations, or RFMOs, discussed further in Chapter 4). These costs are typically paid out of general government revenues. The opposite is true, incidentally, of recreational fishers, who often pay license fees that may be a profit generator for local or regional governments.

Subsidies can not only cover the costs of the local infrastructure needs of a particular fishing fleet, but can sometimes also pay for access to fish in foreign waters. Some governments allow access by foreign fishers to stocks within their EEZs in return for some form of payment. In principle there is nothing wrong with such an arrangement. The countries in question generally do not have the domestic capacity to fish for these stocks, and have need of the funds generated by the arrangement. Pacific island states Kiribati and Tuvalu each earn nearly 42 percent of their GDP from access fees.[9] In some cases, governments also introduce requirements that the fish caught be landed and processed domestically, as a way to promote industrial development. Some Pacific Island states with small populations and large tuna-bearing EEZs have that requirement. Their sale of fishery access rights to foreign fleets, however, often introduces severe management problems, as is discussed in Chapter 4. This access is frequently paid for by the governments of the foreign fishers seeking entry. For example, the European Union buys access to the EEZs of West African countries, paid for out of general revenues. These rights cost the actual fishers nothing. They receive access to a resource for free, while the costs of using that resource (both in financial and environmental terms) are being covered by their governments and by society at large.

Once fish have been caught, governments continue to provide subsidies in ways that can encourage overcapacity. Subsidies to the fish processing industry make fishing of some species in some areas viable when it otherwise would not be. Subsidies in the amount of $700 million in the Asia-Pacific region have helped countries in that region develop local processing facilities that can target local stocks. And subsidies for marketing and promotion help to create consumer demand for fish stocks that might otherwise not be as desirable for fishers to target. The European Union subsidized processing and marketing as part of its Financial Instrument for Fisheries Guidance (which ran initially from 2000 to 2006). Greece used this EU funding for marketing campaigns and packaging development to increase the desirability of Greek-caught sea bass and sea bream in export markets in Northern Europe. Marketing and promotion costs are borne by private actors (manufacturers and retailers) in most industries. Government payments for these activities are therefore really just another form of direct subsidy, as they free up money for operations that fishers or fish processors would otherwise have to pay for marketing.

Beyond payments to fishers that make it cheaper for them to fish, there are also governments that pay fishers not to fish. Fishing is often a seasonal industry, and some countries structure unemployment or social welfare benefits in ways that support fishers in their communities in the off-season. Unemployment compensation and other such support has the effect of keeping fishing communities intact and fishers solvent. An example of this pattern can be found in the Newfoundland fishery in Canada, where fishers often fish just enough weeks per year to qualify for unemployment insurance, and would not be able to remain in the industry, and possibly not in Newfoundland at all, were it not for this form of government support. Unemployment insurance that pertained specifically to fishers in the Atlantic Provinces in Canada was found to actually increase the number of fishers in the region.[10] The European Union provided significant "laying up" grants to fishers affected by moratoria on certain species during the 1990s, and Finland, Germany, and Italy still provide this kind of assistance. Unlike traditional unemployment insurance that is accessible by any wage-earning worker, this type of compensation goes to self-employed fishers who, if they were in any other industry, would not be able to access such governmental support. Compensation for fishers who are unable to work can provide an important social safety net to those who might otherwise face hard times when the fish they target are depleted or regulated. But it also has the effect of keeping more people employed in the industry than would otherwise be. And the more people employed in the industry, the greater the fishing capacity, and the greater the political constituency in favor of government support of maintaining excess capacity.

Some small countries that are both economically reliant on fisheries and have competitive industries subsidize their fisheries only modestly. In Norway, for example, subsidies tend to be worth only about one-tenth of the value of domestic landings of fish. The equivalent figure in New Zealand is only about six percent, and in Iceland less than five percent. The figures for larger countries in which the industry is relatively less economically important, however, are often much larger. Among major fishing countries in the European Union, for example, the equivalent figure ranges from roughly 13 percent in Italy and the United Kingdom to 20 percent or more in France and Spain. The government subsidy is thus worth more than a fifth of the total value of fish landed at the dock. In other countries, the figure is higher still. It is a quarter in Japan, and over one-third in both the United States and Canada. If an industry requires a government subsidy worth more than one-third of its total output, it is probably significantly larger than the state of the resource and the market suggest it should be.

Fisheries as Development

The political power of fishers as an interest group means that governments are prone to subsidizing existing fishing activities to keep them operating. And because fish are essentially free for the taking, governments are prone to using them as a vehicle for economic development. Development lenders such as the World Bank have supported the creation of new fishing industries to take advantage of specific underexploited stocks, adding capacity to an already overcapitalized international fleet.

The World Bank has given considerable development aid to the fisheries sector. Initially, it gave development loans for large-scale industrial vessels and processing facilities. More recently, in light of the global decline in capture fish stocks, it has shifted some of its lending to fishery conservation and development of aquaculture, but it nevertheless still gives loans for the purpose of decreasing poverty through increased fishing. One example of the World Bank approach to fisheries development is a set of three loans to the fisheries sector in the Maldives. The first project (1979–1983) focused on mechanizing the country's fishing fleet. The second (from 1983–1991) focused on infrastructure development to increase the export market for frozen tuna, and the third (1992–1997) involved modernizing the artisanal fleet and providing additional resources for fish processing. The World Bank sees these projects as having achieved the goal of rapid development in the fisheries sector and contribution to national economic growth. More recently, however, the catches per unit effort have declined and there are concerns about overcapacity and overcapitalization in the fleet. In addition, the move away from artisanal vessels, which fish throughout the year, to industrial vessels, which fish only seasonally, has resulted in uneven supplies to processors and erratic employment opportunities.[11]

The use of fisheries as tools of economic development is not confined to developing countries. Developed countries have also been known to use fisheries to promote development in their poorer regions. In Europe as part of the EU Common Fisheries Policy, a European Fisheries Fund exists to, among other things, "help boost economically viable enterprises in the fisheries sector and make operating struc-tures more competitive."[12] Another example can be found in Canadian government attempts to use the Grand Banks cod fishery to promote economic development in its prov-ince of Newfoundland, as described further in Chapter 4. In the process of expanding its regulatory jurisdiction to 200 nautical miles, the Canadian government subsidized the expansion both of the Newfoundland (and to a lesser extent Nova Scotia) fishing fleet and its fish-processing industry. The Grand Banks cod stock that had supported international fish-ing efforts until the 1970s was not able to support Canadian efforts to economically develop Newfoundland. By the early 1990s it was clear that the cod fishery was in crisis, and by 1995 there was a complete moratorium in place on cod fishing in the Grand Banks. With only a few local exceptions, the mor-atorium is still in place – the cod stock has never recovered.

Providing capital for fishing expansion seems at first glance like a viable mechanism for economic development. But because of the nature of the fishing industry, and fish as a resource, this mechanism is inherently self-limiting. The supply of fish in a fishery is capped by biological factors – fish too much, and the stock will decline or collapse. There is, as such, a fixed upper limit to the sustainable yield of any given fishery. In principle, the fishing industry could be developed to the point at which the harvesting capacity of the industry matched the reproductive capacity of the stock. But neither side of the equation is static. Fish stocks are not perfectly stable, and therefore a fishing industry that is of just the right average capacity will over-harvest the fishery in some years and under-harvest in others. And years of harvesting more than the stock can sustain can alter its productivity in future years, sometimes dramatically.

Perhaps more importantly, though, fishing capability is rarely static. A combination of increasing capitalization and

improved technology tends to make each fisher more productive over time. As a general rule of thumb, capacity per fisher
tends to increase by between two and three percent per year.
A fishing fleet that has the capacity to fish a stock at its maximum sustainable yield must shrink its labor pool by two to
three percent per year, overfish its stock, or underemploy its
fishers to an increasing degree over time. As such, fishing
makes an awkward industry through which to promote economic development – it does not matter how successful the
industry is in finding a market, it cannot expand beyond the
fixed constraints of stock size. And the more successful
the industry is at making itself efficient, the more quickly it
needs to decrease its size to avoid overfishing the stocks on
which it depends.

Support of fishing as an engine of economic development
sometimes results in governments subsidizing smaller-scale
or artisanal fishing rather than larger-scale, more industrial
fishing. Such an approach can also result in unintended outcomes. One is that the rules created to distinguish between
small- and large-scale fishers themselves can have a major
impact on the form of the industry. States frequently distinguish between small-scale and large-scale fishing based on the
size of the vessels. Because restrictions on catches from small-
scale vessels are likely to be weaker, fishers have incentives to
build the biggest vessel possible that still qualifies under the
lower-regulation size limit. Rules that favor small-scale vessels for reasons of economic development can, ironically, lead
to larger average vessel sizes in a fishing fleet.

A second unexpected outcome of fishing support as a tool
of economic development is that regions can become even
more economically underdeveloped over time. Subsidies
result in a greater economic focus within a region on the fishing industry than would otherwise be the case, and can attract
new people to the region to take advantage of government

incentives. But the subsidies usually result in fishing capacity surpassing catches fish stocks can support. If this excess capacity results in a decline or collapse of fish stocks, a region that has retooled to support a fishing industry then faces a shrinking or collapsed fishery with more people and fewer other options than if the fishery had not been expanded in the first place.

Flags of Convenience

Widespread government involvement thus has led to serious problems of overcapacity in the global fishing industry. By and large, governments try to support their own fishers and their own fishing industry. But figuring out what counts as one's own industry can be tricky. In particular, governments sometimes end up supporting parts of the fishing industry that they do not regulate. In other words, there are cases where fishers count as national (or local) for the purpose of subsidization, while their vessels count as foreign for the purpose of regulation. In the worst case, these factors combine to create a fishing industry that is at the same time subsidized and unregulated (or under-regulated). The legal mechanism that makes this outcome possible is the open ship registration system.

Any commercial vessel operating in international waters needs to be registered with, and regulated by, a recognized state, and must fly the flag of that state. If a vessel operates mostly within a particular country, it will often end up flying the flag of that country. Many countries, for example, have rules that require that ships that sail from port to port within the country need to be flagged domestically, meaning that they are subject to the full range of shipping regulations (and taxes) of that country. But vessels that operate predominantly, or even frequently, in international waters can generally choose to flag with countries other than those in which they are based, or from which they are owned. When they do so it is often because other flag states provide lower levels of regulation and taxation than their home states do. States that offer lower levels of regulation and taxation in order to attract registry by foreign ships are called flags of convenience.

Fishing vessels may register with flags of convenience for a number of reasons, including lower operating costs (the ability, for instance, to hire workers at wages lower than the minimum wages of their homes states). But the most important reason for a fishing vessel to register with a flag of convenience is to avoid having to abide by quotas and other agreements that constrain its ability to fish. The governments of many of the world's major fishing fleets cooperate in trying to manage global fisheries by participating in regional fisheries management organizations (as discussed in Chapter 4); flag-of-convenience states (which are usually small and poor) often do not. Ships registered in these states are therefore not covered by negotiated quotas, as they would be if they registered in their home states.

Ships registered in flags of convenience have an enormously detrimental effect on high-seas fisheries. More than ten percent of the world's fishing vessels capable of fishing on the open ocean fly flags of convenience, and some put the estimate above 20 percent. That statistic understates the problem, however. Ships that choose this form of registration are likely to be larger than those registered in the home state of their owners. And of the fishing vessels whose registry is "unknown" by the international entities that track ship registrations, most are likely to be flagged with flags of convenience. More importantly, catches of high-value high-seas species by ships flying flags of convenience routinely undermine the ability of regional fisheries management organizations to regulate catches. Depending on the fishery, at least ten percent and up to 50 percent of catches may be taken by unregulated vessels, most of which fly flags of convenience.[13]

The irony here is that fishing vessels flying flags of convenience may end up nonetheless being subsidized by home governments whose regulations they are trying to circumvent. Subsidies targeted either at fishers, such as seasonal unemployment support, or at infrastructure and services, such as port construction or waterway maintenance, support fishing activities, and therefore overcapacity, wherever the vessels themselves are flagged. And governments occasionally even directly subsidize vessels themselves even when those vessels are not flagged in the traditional national manner.

An egregious example of this subsidization is the Atlantic Dawn, the world's largest fishing vessel. It was built in Norway, partially funded by Norwegian subsidies to support its domestic shipbuilding industry, and then registered in Ireland, the nationality of its owner. In this case the ship was not flagged out to a different country, but it was registered as a merchant ship, rather than as a fishing vessel, because registering it as a fishing vessel would have exceeded the authorized size of the Irish fishing fleet under EU rules. But flags of convenience played a central role in this story, as well. The European Union eventually agreed to allow the Atlantic Dawn to be registered as a fishing vessel in Ireland and even to fish in EU waters during part of the year in a one-time deal that increased the allowable size of the Irish fleet by precisely the size of the ship. In return, a different Irish-flagged ship owned by the same owner was transferred to the Panamanian flag and continues to fish on the high seas.[14]

The Market Structure of Fish Consumption

The fishing industry as a whole, then, is incredibly varied, ranging in scale from the purely artisanal to the multi-million dollar, and in range from the purely local to the truly global (trawlers that can stay at sea for months and freeze their catch, and can fish in any of the world's oceans). What happens to the fish once they are caught? The distribution side of the world's market for seafood is every bit as varied as the fishing side.

The vast majority (approximately 77 percent) of global fish production is used for primary human consumption. Of that, about half is consumed fresh, with the rest processed, either by freezing or other forms of preservation. Fish products not used for direct human consumption are used primarily for fish meal and fish oil, which are used mostly as animal feed in agriculture or to feed other fish in aquaculture.

The market for fish is different from most other global commodity supply chains in that there are no individual firms, or any small groups of firms, that dominate an aspect either of the production or the distribution of the product. The petroleum industry, for example, is home to some of the world's largest corporations, and these, along with national oil companies in some of the biggest producer countries, control a majority of the world's production and supply. The agriculture industry is much more diffuse, and a significant proportion of the world's food is either sold locally, or (as subsistence farming) is not sold at all. Even in agriculture, however, a handful of agrotechnology companies dominate the world's market for seeds for staple crops, and another handful of food companies process an ever-increasing proportion of the world's food. There are no direct equivalents to these companies in the global fishing industry.

At the production end of the industry, artisanal fishing (by definition) does not involve large companies. But even at the most industrial end of the spectrum, there are no companies that individually control a significant proportion of global fishing capacity. Iceland has a regulatory system for its fisheries that is more conducive to industry consolidation than most, and capacity there is concentrated in the hands of four major companies. But these companies are big only in an Icelandic, not in a global, context, and industry capacity tends to be less concentrated, often radically less concentrated, elsewhere. Many developed countries, such as the United States and Canada, have regulatory systems designed to promote small operators over industry consolidation. And fishers in developing countries are often closer to the artisanal than the industrial end of the industry. While many industrial fishers operate at a global scale in the sense that their vessels have global range and they fish far from their home ports, there are no companies at the extractive end of the industry that own capacity on a global scale. And there is no equivalent to agro-technology companies in the industry because it is based on capture in the wild.

At the distribution and processing end of the industry there are some larger, and more international, players, although even this end of the industry is less concentrated than is the case in most resource industries. Three major companies (Heinz, Thai Union Frozen Foods, and ConAgra) own the three largest canned tuna operations (Starkist, Chicken of the Sea, and Bumblebee) and thereby can have some degree of collective impact on the market. But even this level of concentration is limited. These companies no longer control their supply chain directly, buying from independent fishers and using independent processing and canning facilities around the globe. In general, the distribution and processing of most kinds of fish remains diffuse and localized. Before discussing industrial and supply chain concentration, therefore, we need to disaggregate fishery-related consumption into a few different industry segments. These range from subsistence fishing, in which the catch never reaches a market, through to a global market for bulk fish protein as an industrial commodity.

In thinking about different industry segments it is useful to think in terms of two distinctions. The first is between localized and globalized consumption. It is impossible to give precise figures for the proportion of the world's fish consumed locally, because the definition of local is imprecise (it can mean anything from the village where the fish is caught, to the country where it is caught). But if one defines the global market as fish products that cross national borders, one can give precise figures for globalized consumption. As noted in the previous chapter, over the past decade, on average just under 40 percent of the world's total fish production has been exported. In other words, about two-fifths of the world's fish production enters the global market, while about three-fifths is consumed in the country in which it is caught or grown (these figures include both capture fisheries and aquaculture). Developed countries export as much as three-quarters of their total production, and developing countries export between a quarter and one-third.[15]

The second distinction is between what can be thought of as the bulk and differentiated segments of the market, the difference between fish protein used as an industrial input and specific species of seafood sold as branded products. Bulk fish goes to a variety of uses, from fish fingers, which (unless they indicate otherwise) generally use the cheapest whitefish available, to fish meal, used as feed for farmed fish or livestock and made from the cheapest fish protein of any kind available. This latter use of fish, for purposes other than human consumption, accounts for roughly 20 percent of total world fish production. At the other end of the spectrum are species (such as bluefin tuna or Chilean sea bass) that one buys or, more likely, orders in a restaurant, by name, and that attract prices in tens of dollars per pound (or tens of euros per kilo) or more.

These two distinctions yield four market categories. Since

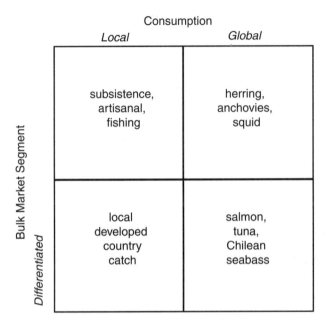

Figure 3.1 Commodity Structure of Fish Products

the distinctions represent ranges rather than points (for example, there are regional markets between the local and the global, and small shrimp are a somewhat differentiated category), not all fish products fit neatly into a single category. But the categories are nonetheless useful guidelines to the market structure of fish consumption. The first of these four categories is the local and bulk. The archetype of this category is the subsistence fisher, who fishes local waters and eats (or shares locally) whatever he or she catches. It also includes much of the fish caught by artisanal fishers and sold at local or regional markets in developing countries, particularly poorer ones, where the fish sold is likely to be that which is not sufficiently valuable to sell to the global market. Local bulk fish found at market are likely to come from artisanal or small-scale fishers. Industrial-scale producers are likely to find it more efficient to sell bulk fish to industrial-scale users (such as fish mealers), whether these users are located locally or globally.

The second of the categories is the local and differentiated. This category consists of fish sold (relatively) locally, but at much higher prices. It includes a large proportion of the catch by smaller vessels in developed countries, where restaurants and other local consumers will pay high prices for fresh seafood. Examples include anything on a restaurant menu prefaced with the term "dayboat." It also includes many shellfish, such as shrimp from the Gulf of Mexico or lobster from Maine. It includes smaller segments of the market in developing countries, where there is less surplus income to spend on designer species. Sources of fish in this category can range from artisanal fishers to industrial trawlers. It can also, counterintuitively, include long-range ships that spend weeks or months between port visits. A fish caught in the high seas is not really local in the normal sense of the term, but still fits into this category if it is a local ship selling its catch on a local market. An example of this process is swordfish caught in the North Atlantic by vessels based in New England (although they may be flagged elsewhere), and sold to restaurants in New York or Boston. This category is much smaller than the previous one by weight, but may be competitive with the category by value, since price per weight can be ten or more times higher for fish sold when the species matters.

The third category is global and differentiated. This category includes fish that are sold by name, as specific species, on the global market. Among these are most salmon and tuna (although not all – some are sold locally), as well as almost all deep-sea fish such as Chilean sea bass (Patagonian toothfish) and turbot (Greenland halibut). It also includes many inshore species – much of the world's shrimp and lobster falls into this category. As is the case with the local and bulk category, sources of fish in this category can range from the artisanal to the large industrial. For example, trawling for Patagonian toothfish in the deep waters off Antarctica can only be done by large factory trawlers, ships that can cost tens or in some cases hundreds of millions of dollars. At the other extreme, some of the shrimp that make it to the US market from India are caught in very small quantities by fishers using little more than canoes with tiny outboard engines. These shrimp catches are aggregated at the village level by distributors who then sell the shrimp on to large-scale exporters. Once caught, fish in this category can end up being distributed by large-scale exporters and wholesalers, as is the case with the shrimp example, or distributed through markets made up of small-scale sellers and buyers, as is the case with the fish that pass through the Tsukiji fish market, considered the world's largest, in Tokyo. Nearly 600,000 tonnes of fish are auctioned at this wholesale market annually, at a total value of approximately $5.5 billion (US).

The fourth and final category in which the market for fish can be divided is global and bulk. This end of the global industry is the most dominated by multinational corporations. Fish that find their way into the undifferentiated market include small mid-ocean fish such as herring or anchovies, bottom-feeders such as pollock, and low-value non-fish seafood such as squid. These fish are likely to be caught at an industrial scale, because aggregating small catches of low-value fish for the international market is less likely to make economic sense for exporters than aggregating small catches of high-value species does. McDonald's, for example, buys 50,000 tonnes of fish annually for its fish sandwiches; it has traditionally bought the least expensive whitefish available, switching ingredients as stock declines or regulations affect certain species. More recently it has attempted to purchase its fish from fisheries considered to be sustainable, as discussed further in Chapter 6. The corporate heft of a business like McDonald's allows centralized decisions to have a broader impact throughout the supply chain.

Unlike most of the local undifferentiated catch, much of the fish in this category is used for non-human consumption (which accounts for just under one-quarter of the world fish catch), mostly fish meal that goes into the global market for agricultural and aquaculture feed. Most of the rest goes into industrial food processing, yielding fish-based industrial food products of the sort one might find in a supermarket freezer or in a fast-food restaurant. The most common product of this sort is a puree of whitefish that is used to make fish sticks and imitation crab.

To this point several sorts of distribution for the global fishing industry have been mentioned. Unlike most resources, the distribution of much of the world's fish catch happens on a small and local scale. Fish are like agriculture in this way. But also like agriculture, much also enters a global supply chain. Some parts of this supply chain work through integrated corporate structures, while other parts are organized into market-based distribution.

The final link in the chain between fisher and eater of fish is the retailer, which serves as the point of direct contact with the consumer. For non-locally consumed fish, the retailer is usually a restaurant or a store. As is the case with all of the other elements in the fish and seafood distribution industry, this point of contact can vary widely in scale, from a small restaurant or fishmonger to a multi-billion dollar restaurant conglomerate or supermarket chain. It is at this stage that consumer choice can impact the behavior of other actors in the industry, particularly those catching fish. By choosing to avoid certain species, or only to buy fish that are caught in a sustainable way, consumers can send signals back up the supply chain that affect both what fishers catch, and how they catch it. Other choices, such as buying the least expensive or best-looking fish, send other signals. These choices, and their effects, are discussed further in Chapter 6.

Conclusion

Fish, as a commodity, faces a complicated structure of extraction and consumption. Fish are caught and marketed through a largely diffuse structure, with fishers across many different scales catching fish for many different types of markets and distribution processes. Fish capture ranges from a purely local activity undertaken to feed a family or a local community, to a large-scale global industrial enterprise seeking undifferentiated inputs. Changing behavior across these numerous and extremely different actors, to protect dwindling fish stocks, can be difficult.

At the same time, many governments take the interests of those catching fish more seriously than their economic importance to specific countries can account for. Because of the political power of those who catch fish, governments subsidize fishing operations and fishing communities, thereby keeping people employed in the fishing industry, even when the health of fish stocks would, in a perfectly free market, lead them to pursue other forms of employment. Governments and development organizations also use fishing as a means of economic development, thus increasing the number of people who come to demand an unfettered right to fish.

This subsidization happens even as global fish stocks can no longer support fishing without limits, in the context of a resource that is particularly sensitive to overexploitation. Because fish are a common pool resource, market forces will not adequately regulate fishing – the industry needs enforced rules to prevent overfishing. The governments that are subsidizing the industry are the same ones that need to be creating and enforcing these rules. The next chapter explores their efforts to protect fish stocks.

CHAPTER FOUR

Regulatory Efforts and Impacts

Although fish are a renewable resource, capture fisheries are in some sense finite. Fish cannot expand beyond their ecological niche, and if they are harvested beyond the point of sustainability their availability declines. And because fish are a common pool resource, it can be difficult to prevent access to them by anyone who might want to harvest them. Even if individual fishers would be willing to use short-term restraint to protect the long-term health of a fish stock, collective action problems and open access to the fish stock render such individual choices unlikely. Those seeking to protect the long-term sustainability of fish stocks have therefore turned to a variety of forms of regulation, with mixed levels of success.

Widespread regulation of fishing in order to reduce the pressure on stocks is a relatively recent phenomenon. Until the era of industrial fishing, the very idea that humans could threaten the bounty of the sea seemed absurd. As recently as the 1950s, despite clear evidence of decline in some major fisheries, the argument was still made that human efforts were not a threat to global fish stocks. A 1954 book written by two marine scientists was entitled *The Inexhaustible Sea*,[1] indicating how even scholars conceptualized ocean resources not much more than a half century ago. Serious international efforts to meaningfully manage global fisheries did not develop real momentum until the 1970s, when evidence of depletion of major stocks could no longer be ignored.

This chapter examines the history of national and

international regulation of fishing from collective action ini-
tiatives undertaken directly by fishers, to rules passed by
governments and international organizations to put limits
on fishing. It also examines other more recent regulatory
options, such as closing some areas to fishing entirely, and
mechanisms to give individual fishers or fishing companies
some form of ownership rights to fish resources. Most of
these regulatory processes face difficulties, for many of the
same reasons that fish are overharvested in the first place. But
the willingness of governments, international organizations,
and fishers to adapt their conservation strategies based on the
results of earlier efforts may lead to a better understanding of
how to manage fisheries for sustainability.

Local Fishing Conservation

There are reasons to expect that regulation of fishing should
happen first at the local scale, where there are small numbers
of fishers involved, they can directly see the results of their
actions, and they are unlikely to have economic options other
than fishing if they fish out their local stocks. Regulation
should also be more likely with relatively stationary species,
like lobsters or reef-dwelling fish, than with migratory species.
In the latter case cooperation may be more difficult because
any community that practices restraint knows that the fish
might not be protected elsewhere along their journey; conservation may therefore be easier for marine species that do not
travel far. Cooperation at the local scale need not be formal,
as is regulation at the national or international level. It can
be unofficial, based on social norms and communal enforcement. Since this sort of cooperation often leaves no written
records, we do not know its full history. But two examples of
recent communal management illustrate the practice.

Lobsters in northern New England used to be so plentiful they were used for fertilizer, bait, or food for people who could not afford better options. With expanding markets, improvement in ways to protect and ship live lobsters further from where they were caught, as well as the development of a lobster-canning industry, they grew rarer and more expensive. What is notable in the lobster fishery in this region is the extent to which those who fish for lobster have been able to self-organize to protect stocks, which have remained stable for more than the last half century. A governmental regulatory structure exists around the self-organized process, the centerpiece of the successful lobster conservation efforts. A core feature of the process is preventing new people from entering the fishery; "gangs" of lobster fishers who all know each other operate in a given harbor or island. They can limit entry of others into the area, often by informal measures to harass newcomers, and have been able to set and collectively enforce catch limits, and ensure that rules about egg-bearing or undersize lobsters are upheld.[2]

The second example is Mawelle, a fishing village in Sri Lanka, which for a time managed a cooperative beach seining fishing operation. The nets used in this fishery were a mile long and more expensive than a single household could support or set. Each net was divided into eight shares, with eight people to work it. The beach conditions are such that only two can be set at any one time, one in the bay (which has bigger but more variable catches) and one in the harbor (where smaller but more consistent catches are likely); the time of day at which nets are set also influences the level of catch, as does the time of season. The villagers devised and enforced a complex rotation of net placement so that the location and time of net setting varied. This system worked as long as the number of nets allowed was limited. But population growth and construction of a road that opened up new fish markets led to pressure to increase the number of nets, and once this system to limit entry by new fishers broke down, the self-regulation eventually failed.[3]

In both of these examples, successful management of the fishery depends on some form of limited entry. This approach takes something that was previously common property (other things being equal, anyone can go out and fish), and closes it so that only a select group can access it. As such, it is a form of what economists call privatization. This approach is easy to implement with farmed animals – they can be branded or tagged, and enclosed on private land. Enclosure is possible with fisheries, but unlikely with anything larger than a pond. So fisheries regulation often tries to come up with other things that fishers can own individually, that make it more likely that they will refrain from overfishing. In the lobster case, it is access to an area where lobsters are to be found. In the Sri Lankan fishery it was shares in a limited number of nets and participation in a rotation scheme. In other cases it may be a license to fish, or in more recent national-scale regulatory efforts, a quota allocation that can be bought or sold, in effect the continuing right to a certain proportion of a fishery. Privatization as a concept also underlies the move to enclose more of the ocean under state regulatory control, discussed below.

What do fisheries management efforts regulate, beyond attempts at privatization? Many forms of regulation have some characteristics of privatization, giving people rights to fish, but only some people, or only a certain amount or type of fish. The licensing of fishers is a form of regulation, that can be designed either primarily to generate revenue for the government (probably the driving logic behind most early forms of licensing) or to limit the number of fishers, and by extension therefore hopefully the amount fished. Fishers can also be given quotas, limits on the amount (either in total, or of specific species) they can fish. Quotas are a key output of international fisheries cooperation, in which a total allowable catch is set; that amount is frequently divided into country quotas, and the countries then determine how to allocate their quotas among their own fishers, sometimes by giving them individual quotas.

But not all fisheries management involves privatization. Regulatory efforts often generate rules about what fishers cannot do. In particular, the rules can limit when, where, and for what fishers can fish, and the equipment they can use. Regulations can limit fishing to specific areas (or put specific areas of water off limits), can limit fishing seasons, and can put certain species off-limits, or require minimum or maximum sizes that fish must be to be taken. They can also prohibit the use of some kinds of equipment, like purse-seine nets or explosives, and put parameters on other kinds, for example by requiring larger mesh sizes in nets than fishers would otherwise use. All of these forms of regulation are used in fishery management efforts, although quotas and restricted areas are becoming more prominent over time as season and equipment regulations prove inadequate by themselves to stem overfishing.

Early Fishing Regulation

Although local cooperation to manage fisheries has a long history, national and international efforts are relatively recent. Until less than a century ago, involvement in fisheries by national governments was focused more on generating revenue or on developing greater national fishing capacity than on preventing overfishing. The perception was that the bounty of the sea was limitless. But by the beginning of the twentieth century, and in some places earlier, the effects of overfishing in some areas and of some species had become clear.

The first major national efforts to regulate fishing tended

to focus on preventing catches of immature fish, as a way to address declining fish stocks. Such regulation usually involved requiring that mesh sizes in nets exceed a certain minimum. For instance, the UK in 1843 required nets used to catch cod to have a mesh size of at least 3.5 inches, to allow juvenile fish to escape. This approach was the first type of fishing regulation followed by international regulatory bodies as well.[4] A next strategy was to declare fishing seasons – specific times when fish could or could not be harvested. This approach can be used to close spawning grounds during times when fish are reproducing and young fish face the most danger, or to end fishing when a large number of fish have been caught.

At the international level, cooperation among states to regulate fishing started relatively early in the history of global multilateral cooperation, but developed slowly. The Convention on North Sea Overfishing of 1882 was negotiated to harmonize domestic fishing regulations among the United Kingdom, Germany, Denmark, Netherlands, Belgium, and France. There was also some cooperation to manage specific fisheries, such as the Fraser River Convention of 1930 between Canada and the United States, which divided the catch of Pacific Salmon from that river between the two countries. Other international agreements that serve as the contemporary basis for international fisheries management were negotiated after the Second World War. These include the International Convention for the Regulation of Whaling in 1946 (although whales are marine mammals, they are regulated as a fishery), the Agreement for the Establishment of the Asia-Pacific Fishery Commission in 1948, and the International Convention for the Northwest Atlantic Fisheries in 1949. These institutions cover either specific regions (such as the Northwest Atlantic or the Antarctic Seas) or types of species (whales, tunas). By the early 1980s this network of institutions, called Regional Fisheries Management

Organizations, or RFMOs, was largely complete, and covered most international waters.

By the 1970s and 1980s, however, it was becoming clear that the system of national management of fisheries within territorial waters (which traditionally extended three miles from shore, although by the 1970s many countries were claiming 12 miles), combined with multilateral management of fisheries in the high seas, failed to deliver effective management of global fisheries. Increasing technological sophistication and growth in the size of fishing operations, combined with the common pool resource nature of ocean fish, led to evident declines in fish stocks among the commercially desirable fish.

Exclusive Economic Zones and the Partitioning of the Global Commons

The biggest change in the approach to dealing with ocean fisheries was the extension, codified in the 1980s, of national control over ocean fisheries that were previously considered to be part of the open ocean. This expanded national jurisdiction brought, by some measures, more than 90 percent of commercially caught global fish stocks under some level of national control. This expansion changed the face of both international and domestic fishing practices and regulatory processes. From the perspective of economic theory, this expansion of national control is a kind of privatization – it made most fish stocks national property, rather than part of the international commons. But, in practice, it ultimately led to a greater intensity of fishing.

Historically, dating back to the 1500s, states had claimed national control over areas of the ocean that extended three miles from their coastlines. The area of the open ocean beyond that distance was considered the "high seas," where the governing principle was freedom. The drive to extend national jurisdiction over fisheries began in the 1940s, when Chile and Peru claimed jurisdiction over the area 200 nautical miles from their coastlines, taking their inspiration from two proclamations by US President Truman. The first of these claimed for the United States the right to explore for oil and gas in the continental shelf that continues beyond the territorial sea. A second proclamation indicated that "in view of the pressing need for conservation and protection of fishery resources" the United States considered it acceptable "to establish conservation zones in those areas of the high seas contiguous to the coasts of the United States wherein fishing activities have been or in the future may be developed and maintained on a substantial scale."[5]

Shortly after the declarations by Chile and Peru, additional states declared 200-mile maritime zones; these ranged from Arab states concerned about access to oil resources on the continental shelf, to Ecuador, which joined other Latin American states in declaring unilateral access to marine resources. Increasing numbers of states declared their own maritime zones. By the time international negotiations began in 1973 toward the United Nations Convention on the Law of the Sea (UNCLOS) many states had declared some form of expanded national jurisdiction over maritime resources, although the basis of these principles, and the actual nature of the zones, were sometimes contradictory or competing.

UNCLOS, negotiated between 1973 and 1982, resolved the issue internationally by codifying the idea of an exclusive economic zone (EEZ). This zone extends up to 200 miles from shore, in which states are authorized to exercise exclusive control over natural resources. National sovereignty still extends no more than 12 miles out, and the area of EEZs still counts as international waters from the perspective of navigation, national defense, and laws not relating to economic resources. Beyond the 12-mile limit of territorial seas but within the EEZ, however, a state has the authority to control and regulate the extraction of resources, whether these resources be on, in, or under the sea. This jurisdiction includes authority over all fish resources, meaning that from the perspective of fisheries management the EEZ can be seen as essentially the same as territorial waters.

One immediate effect of the expansion of national jurisdiction over fisheries resources was the impact on international regulatory processes. Fish caught on the open ocean have, if managed at all, been regulated by a set of international organizations, discussed in greater detail below. These organizations had traditionally managed fisheries beginning either three or twelve miles out from the shoreline, at the end of what states considered to be their national waters. In the wake of UNCLOS many of the agreements creating these organizations were either amended, or renegotiated, to take account of the new global arrangements on ocean jurisdiction. Not all were, however; international tuna agreements, for instance, still regulate tuna catches wherever the relevant species are found, as is true for whales governed by the International Whaling Commission.

The other difficulty that arose was how to deal with fish stocks that crossed between different areas of national jurisdiction, or between national and international waters. UNCLOS indicated that states with straddling stocks in their EEZs should cooperate with each other and with the relevant regional fisheries management organization to manage these stocks, but it soon became clear (in part when Canadian Coast Guard vessels impounded a Spanish vessel fishing just outside of the Canadian EEZ) that these fish species would be the source of international contention.

International negotiations to address the problem of straddling and highly migratory fish stocks had been underway since 1992, and in the wake of the "turbot war," as the

Canada–Spain conflict was known, an agreement was reached. The resulting agreement bears the unwieldy name of *The Agreement for the Implementation of the Provisions of the United Nations Convention on the Law of the Sea Relating to the Conservation and Management of Straddling Fish Stocks and Highly Migratory Fish Stocks*, and was agreed to in 1995. (It is more commonly called the Straddling Stocks, or Fish Stocks, Agreement.)

Post-EEZ Behavior

There were important additional effects of taking large portions of the ocean out of the management control of international organizations and handing those areas to the jurisdiction of states. UNCLOS gave states rights over natural resources for the purposes of "exploring and exploiting, conserving and managing the natural resources, whether living or non-living."[6] States had the opportunity to manage or exploit the stocks they now controlled as they saw fit. Most states with large EEZs applied some combination of national regulatory measures with domestic subsidies to increase the intensity with which their own fishers were able to exploit these newly privatized resources.

States could – and most initially did – require foreign fleets that had been fishing in these newly national areas to leave. In some cases foreign fleets were allowed back in, but only after the negotiation of access agreements, that either involved reciprocal rights to fish in each other's waters, or specific payments for access to fish (often, but not always, with limits). This issue is discussed in greater detail below.

In theory, national regulation could have mitigated the "tragedy of the commons" that occurred when fishers from many nationalities were competing for a limited resource. Enclosing the resource to allow management control by one country could have resulted in the imposition of strict standards and the protection of stocks, especially since national governments have enforcement authority that is lacking in international management structures. Such privitization of resources is often suggested as one type of solution to resource management problems, and was a key element of the discussion at the UNCLOS negotiation. If international management had not succeeded in protecting ocean fisheries, perhaps national management could. A frequent state reaction to newly created EEZs, however, was to subsidize the domestic fishing industry (as discussed in Chapter 3), thereby further exacerbating the global problem of overfishing.

In many cases, countries substituted a domestic tragedy of the commons for an international one. Governments interpreted their new control over fisheries as an opportunity to generate economic development, by creating the industrial capacity to take advantage of exclusive national access to fish stocks in their EEZs. The creation of new industrial capacity (both fishing capacity and onshore capacity, such as port facilities and fish processing plants) was encouraged through the use of subsidies, although in centrally planned economies governments built capacity directly. Unfortunately, the enthusiasm for economic development often led to capacity overtaking what the fisheries could support. Because of subsidies, increasing numbers of fishers began to fish in these new areas of domestic control, often adding to the pressure on the newly nationalized fisheries, and occasionally leading directly to the collapse of fish stocks.

An example of many of these processes in action can be found in the cod fishery on the Grand Banks off the Atlantic coast of Canada. The Grand Banks is an area of continental shelf that extends from the southeast coast of the Canadian province of Newfoundland. Most of the area is now located in Canada's EEZ, but until the creation of that EEZ the

Grand Banks were almost entirely in international waters. They have long been known as one of the world's richest fishing grounds, and are one of the oldest known major long-distance fisheries. The Vikings fished there a millennium ago, and the Basques, from what is now northern Spain, began fishing there perhaps 800 years ago. Europeans (as well as, more recently, people from New England and Canada's maritime provinces) fished there on a large scale until Canada's declaration of its EEZ in 1977. This fishery was essentially unregulated (the International Commission of the Northwest Atlantic Fisheries, a precursor to the Northwest Atlantic Fisheries Organization, collected data but did not have the power to set quotas). It was probably overfished, but not critically so.

Newfoundland has long been one of the poorest parts of Canada, and the federal government saw its new control over the Grand Banks fisheries as an opportunity for economic development there. It began by reserving the area for Canadian fishers, excluding the foreign fleets that had traditionally fished there. Fishing had long been a mainstay of the Newfoundland economy, but the local industry did not have the capacity to replace the European and American fleets that were newly excluded, so the Canadian government undertook to subsidize a major expansion of the industry in Newfoundland. A key difference between the Newfoundland fleet and the foreign fleets it replaced was that the European and American ships had much longer range. When excluded from Canadian waters, they could fish elsewhere. The Newfoundland fleet, designed to fish within 200 miles of shore, was much shorter range, and therefore much more dependent on its home waters. When the cod ran out, most Newfoundland fishers had nowhere else they could go.

The Newfoundland fleet was relatively short-range in part because the government wanted to promote small-scale, locally owned fishing, rather than large industrial fleets. Its research showed that most fishing vessels in Newfoundland at the time fell into one of two categories, small boats in the 30–40 foot range, and a much smaller number of bigger factory trawlers in the 100-foot range. The new policy split the difference, subsidizing vessels of under 65 feet, but not bigger ships. The result of this arbitrary distinction was that most new vessels built in response to the subsidies were between 64 and 65 feet long.[7] A policy intended to promote the use of smaller vessels had the unintended consequence of making the median Newfoundland fishing vessel almost twice as long as it had previously been. Fishing capacity thereby expanded much more rapidly than had been envisioned.

The results of this rapid expansion were as would be expected – more fishing. Some scientists did warn of impending doom if catches were not capped, but they were largely ignored. Because Newfoundland fishers had nowhere else to fish, and to pay for the big new vessels they had bought (which were subsidized, but not free), there was strong opposition within the province to major reductions in the amount that could be fished. And because better management of the fishery would lose votes in Newfoundland but not win them anywhere else, there was no political incentive for better management. There was also no precedent for serious concern – a fishery as rich as the Grand Banks had never completely collapsed before, so scientific arguments about the dangers generally fell on incredulous ears. Furthermore, in the few years between Canada's declaration of its EEZ and the growth of the new Newfoundland fleet, stocks of cod had recovered somewhat from overfishing in the 1960s and early 1970s, making catches seem quite healthy for a while.

By the mid-1980s there was clear evidence that stocks were under stress, but not enough was done to restrain fishing activity, and by the early 1990s stocks had collapsed almost completely. At that point, there not being enough fish left to support a commercially viable fishery anyway, the Canadian government declared a total moratorium on cod fishing within its Atlantic EEZ. That moratorium remains mostly in place today, but after two decades of moratorium, the cod have for the most part not recovered, and there is a real fear that they never will. In the meantime, the Canadian government, having paid to build the fishing fleet in the first place, has now spent several billion dollars both buying back and destroying those same vessels to reduce capacity, and supporting and retraining fishers who can no longer fish.

Developing countries followed similar paths, especially newly industrializing states. Chile had been one of the first states to push the idea of national control of an expanded area of ocean, and after such an approach was codified in the Law of the Sea, Chile moved to take advantage of its new owner-ship of resources. It excluded (or required licenses to be bought by) foreign fishing vessels in the area, and replaced the requirement for domestic fishers to have fishing permits with a completely open access system in which anyone could fish. Chile had already been liberalizing its trade, and a state export promotion agency worked to increase the marketability of Chilean fish (including the newly branded Chilean sea bass) in export markets. Chilean catches grew 8.6 percent per year through the 1980s, until clear overcapitalization and overfish-ing finally led the Chilean government to impose quotas and minimum size requirements in some of the main fisheries in the area.[8]

Domestic Fishery Regulation post EEZ
The creation of EEZs brought most of the world's fishing resources within the purview of national regulators. National governments can choose to close their EEZs to all but domes-tic fishers, or they can sell fishing rights to foreign vessels.

(They could in principle give away the right to fish to foreign vessels, but in practice this approach is understandably rare.) As the Grand Banks cod case shows, this nationalization of much of the ocean's bounty does not necessarily lead to better management. When governments see their control of EEZs as an opportunity for economic development they are prone to oversubsidizing their fishing industries, leading to greater capacity fishing those particular waters than would have been the case otherwise.

Selling the right to fish in an EEZ to foreign vessels can also be problematic. It need not be so – a national fisheries regulator can determine what level of exploitation fish stocks within its EEZ can bear, sell the right to fish only that much, and police the result. But in practice the selling of fishing rights often leads to overfishing, for two reasons. First, since governments who sell fishing rights to foreigners are generating cash payments rather than supporting a domestic industry (with domestic constituents who might care about the long-term health of the resource), there can be an incentive to generate as much cash in the short term as possible. States facing fiscal problems, and states with corruption issues, where the money generated gets siphoned off by officials rather than being used for public policy purposes, are particularly prone to selling more access than the stock conditions warrant. Because larger, wealthier countries tend to have established fishing industries fully capable of exploiting local resources, the countries that sell access to their EEZs to foreign vessels tend to be smaller and poorer countries, which in turn are more likely to have governance difficulties.

The second reason that selling fishing rights internationally can undermine effective management of fishery resources is that the countries selling these rights often do not have the capability to effectively police their EEZs. One example of this problem can be seen in the small island states of the Pacific

Ocean, such as Nauru, the Solomon Islands, and Vanuatu. These are countries with populations of a few hundred thousand relatively poor people, but with EEZs covering hundreds of thousands of square miles. They simply do not have the resources to police this territory effectively. West African states, which often sell fishing rights to the European Union, also rarely have the resources to monitor the agreements they set up. The access agreements with the EU issue licenses for a specific type of fishing vessel, based on what vessel owners say they intend to catch. But there is little oversight about what these boats actually do catch. For instance, only eight percent of the catch of Italian-flagged vessels with shrimp licenses in the waters of Guinea-Bissau in 1996 was shrimp; most was other types of fish and cephalopods (such as squid or octopus). Portuguese-registered shrimp vessels, similarly, had catches that were only 27 percent shrimp.[9] And the fact that the European Union is paying for the right to fish, rather than the European fishers themselves paying, is another form of subsidy of the EU fishing fleet.

A particularly pernicious example of the failure of governments to effectively police their own EEZs can be found in Somalia. It is a failed state; there has over the past two decades been no real Somali government to enforce any rules about fishing in its EEZ. As a result, illegal fishing became rampant – anybody could fish there, without paying for the privilege and without limits. The predictable result was that foreign fleets were overfishing, leaving little for local Somali fishers. The local fishers responded by arming themselves and organizing into groups to physically chase off illegal foreign fishers. Once they had learned how to be effective at chasing down ships, they turned to more lucrative pursuits than fishing. It is these local groups that eventually became the Somali pirates that have been preying on shipping in the waters off the Horn of Africa for the last several years.[10]

Most states nevertheless make some efforts to manage their fisheries, and a few succeed in managing them well. Fisheries are generally managed by a specific unit of the national government, although in federal systems jurisdiction can be complicated. For example, in the United States state-level governments have the authority to regulate fishing in inland bodies of water and in territorial waters, out to the 12-mile limit, while the federal government has the authority to regulate fishing in the EEZ. The situation is similarly complicated in the European Union. Regulations are made at the EU level, but enforcement is the responsibility of individual states. This bifurcation can create a mismatch between a desire by the European Commission to manage EU waters as a single entity and desires at the national government level to protect national fishing industries even if at the expense of other EU members. The result is often a classic collective action problem – inadequate enforcement at the national level, because the costs of overfishing are shared among all who want to access the stock, but the benefits in the short term of catching the fish accrue to those who do it. Meanwhile, fishing may be subsidized by multiple levels of government, in both federal systems and in the EU.

Regulators are generally responsible for the fishing industry – their remit is basically socioeconomic, rather than environmental. In other words, their job is to keep the industry healthy, not to protect fish as such. In the long term, keeping the industry healthy means preventing overfishing, but in the short term regulators' responsibility to the industry can create mixed incentives, because regulation imposes costs on fishers, and quotas and other mechanisms to limit fishing effort have the effect of putting fishers out of work. The pressure on regulators to respond to the short-term needs of the industry is in many cases exacerbated by increasing demands for community involvement in the management of fisheries, and for increased responsiveness to the social needs of fishing communities as well as the economic needs of the industry. This pressure can result in policy that is better informed by social conditions and more responsive to local needs. But it can also result in pressure to maintain the economic health of the industry, even when fishing stocks are inadequate to support it at its current size.

From the perspective of fisheries regulators, the best result is to keep all the fishers it regulates in business while limiting the amount fished, by getting resources from the rest of the economy and transferring them to fishers. This regulatory incentive contributes to the heavy subsidization of the industry globally. Governments pay fishers not to fish, or to fish elsewhere, keeping more people in the fishing industry than fish stocks can support. Subsidies in turn exacerbate the vicious cycle, by keeping too many underemployed fishers in the industry, who in turn demand continued subsidies.

Given this incentive, regulation tends to work best in one of two circumstances. The first is when the damage to stocks from overfishing is so obvious that it can no longer be ignored. In the Grand Banks example the crash of the cod stock was so complete that neither the government nor the industry could deny that it was a disaster that required radical action. Even extreme overexploitation, however, will not necessarily lead to sound management. It is generally accepted, for example, that the bluefin tuna fishery in the eastern Atlantic Ocean and the Mediterranean Sea is on its way to environmental disaster. But the governments most responsible for the fishery continue to prevent an adequate regulatory response.

The second circumstance is when the fishing industry is a sufficiently important part of the national economy that the government cannot afford to subsidize it heavily without placing an undue burden on the rest of the economy. These conditions apply in small island nations with tiny populations, such as those in the Pacific tuna fishing belt, and Iceland. For other countries with small populations and large coastlines, such as New Zealand and Norway, large-scale subsidies of their fishing populations would require relatively large expenditures. The poorer of these countries have trouble policing, and therefore effectively managing, their EEZs. But the wealthier ones, such as Iceland, Norway, and New Zealand, have among the world's best-managed fisheries.

International Regulation – Regional Fisheries Management Organizations
Despite the adoption by most countries of EEZs, some of the world's fisheries (including some of the most lucrative ones) cannot be managed by countries individually. Some fish live in what remains the high seas, beyond the 200-mile limit, and others either move between the high seas and national waters, or regularly swim between national jurisdictions. Because there is no international authority able to govern fishers directly, these stocks can only be managed effectively if all the countries whose vessels fish the stocks cooperate. This cooperation is generally done through the creation of international organizations, called regional fisheries management organizations (RFMOs). These organizations are entities created by states, and run by states, which vote on the regulatory measures they impose. A key limitation of international organizations is that membership in them is voluntary – no country has to join. It is difficult to persuade countries to participate in responsible management of international fisheries if they (or the fishing vessels they register) prefer to stay outside an organization and not be bound by its rules.

These organizations regulate fishing behavior in the context of a geographic area, a fish species, or some combination of both. Although each RFMO operates separately, they follow broadly similar procedures. They generally regulate fishing behavior by passing rules pertaining to catch seasons, catch amounts, and acceptable fishing methods. As with domestic fishing regulations, international regulation began with gear restrictions, particularly a focus on ensuring that juvenile fish could escape from nets. But that approach is insufficient if too many fish overall are taken, and international regulatory approaches quickly focused on setting quotas on catch volumes.

Initial quotas, as in early international efforts to regulate whaling, Pacific tuna, and Pacific halibut, were set collectively; a total catch limit applied to the fishery without specifying who was allowed to catch it. A season ended once the full allocation had been caught, regardless of who caught what amount. This approach led to a "race for the fish" and to overcapitalization, because the amount any state, or individual fishing vessel, could catch was negatively affected if others caught fish first. Those who could catch quickly, therefore, could gain a greater share of the overall catch. Overcapitalization is likely in that situation, because faster ships, with better technology, can find and capture fish more quickly. (It also, likely, led to misreported catches, as the more anyone caught, the closer the season was to ending.) By the 1960s international rules instead set national quotas as parts of an overall total allowable catch (although how, and whether, those were further allocated depended on the individual country). These organizations also often regulate fishing activity by restricting certain types of gear or closing seasons or areas to fishing to protect the most vulnerable stocks.

In most cases, the primary decisionmaking body in an RFMO is a Commission, in which each member state is represented and has one vote on policies. Fisheries commissions usually meet annually. Before decisions are made, recommendations are often made by a Scientific Commission, that either conducts its own scientific research or, more frequently, bases its estimates on work done by others. The Scientific Commission does not have decisionmaking power, but estimates the status of the relevant fish stocks and the likely maximum sustainable catch.

Potential policies, informed by the Scientific Commission, are then put to a vote in the Commission. Although some fisheries commissions, especially those created recently, require unanimous voting to pass rules, many are empowered to pass regulations by a specified supermajority rather than unanimity. These commissions, however, also include "reservations" procedures, by which states that are outvoted in the passing of regulations can choose to avoid being bound by those rules. Such procedures generally also include a two-stage process, so that if one or more states choose to opt out of the rules, additional states can, in response, also avoid being bound by them. No state wants to find itself in a situation in which it is bound by rules that others are able to avoid.

The effects of such reservations processes are complex. On the one hand, there have been fishing seasons in which many or most relevant states opt out of rules passed by a fishery commission, clearly undermining efforts at protecting stocks. For example, in 1995 in the Northwest Atlantic Fisheries Organization (NAFO) the European Union, along with Estonia, Latvia, Lithuania, and Poland, all objected to the allocation of quotas for Greenland halibut. In 1957 in the International Whaling Commission (IWC) the Netherlands, United Kingdom, Panama, South Africa, Norway, Japan, the United States, and Canada (the primary whaling states at the time) all eventually objected to the reduction in whale catches passed that season.[11]

On the other hand, these reservations do not create disagreement over management decisions; they simply reflect it. A requirement for unanimity in fishery commissions – as has been the trend in more recent agreements – would simply prevent rules that some states oppose from being passed at all. In the best of circumstances, reservations procedures allow a subset of states willing to be regulated to go ahead without the cooperation of a reluctant state, and ideally demonstrate to others that regulation is either necessary or not too onerous. In some cases (most notably in the International Whaling Commission) states with reservations to regulations have been persuaded to remove those reservations and join the regulatory process.

Another unusual element of international fishing regulation is that catch limits are generally decided annually (although sometimes in the context of multi-year plans). Although annual catch limits require new political negotiations every year, setting them annually allows regulations to respond quickly to new scientific information about stock abundance and to adapt to specific conditions in the fishery. For example, the "turbot war" mentioned above began when the Northwest Atlantic Fisheries Organization, in response to new evidence about rapid stock depletion, recommended a new quota of 27,000 tonnes of the fish for 1995, less than half as much as had been caught the previous year.

The regulatory process in RFMOs allows for annual decisions to take account of new scientific information about the health of a fishery and respond accordingly. In practice, however, political jockeying means that RFMOs frequently fail to protect the fish species they manage. Even in commissions where unanimous voting is not required, states know that policies that are opposed by the major fishing states are unlikely to be implemented. And the fishing states face their own domestic pressure to refrain from undertaking short term measures that would hurt domestic industry, even if these measures would have the long-run effect of protecting the fishery and would thereby benefit domestic fishers eventually. For that reason, commissions regularly agree to policies that are more lenient, or involve higher catch limits, than recommended by Scientific Commissions. The International Convention for the Conservation of Atlantic Tunas (ICCAT), for instance, routinely passes quotas that are considerably higher than its scientific commission proposes.

RFMOs face other difficulties once their regulations are in place. Compliance poses a difficult problem for international organizations generally. Because there is no world government to enforce international law, states must cooperatively set up any monitoring or enforcement mechanisms along with whatever cooperative measures they create. Monitoring is difficult and enforcement even harder, so it can be hard to ensure that states are upholding the rules they have agreed to. The ocean is vast and individual fishing vessels are numerous; it is difficult to determine what any one of them is doing at any point in time, and fairly easy for anyone who wants to evade the rules to do so.

Some non-compliance exists at the margin: fishers might mis-report catch sizes so that catches of fish just below the legal size limit are reported as just above. Early whaling statistics provide an example. In 1965 when the minimum size limit for baleen whales was 38 feet, 90 percent of female baleen whales caught were reported to be between 38 and 39 feet long, a statistical impossibility. Whalers were almost certainly catching undersize whales and reporting them at or near the minimum size. Or it may be catch location that is mis-reported. Once catches of Patagonian toothfish (Chilean sea bass) were limited by the Commission for the Conservation of Antarctic Marine Living Resources (CCAMLR), an increasing number of vessels reported their catches were taking place just outside of the CCAMLR regulatory area, accounting for more catches in that region than would have been biologically possible. Although statistical analysis of these catches may turn up improbable size distributions or location reports and thereby lead to speculation of illegal catches, that kind of analysis is rarely done by the organization itself and can almost never lead to pinpointing specific fishers that have caught below-size or otherwise questionably legal fish.

Other non-compliance is more devious. Because there are few points at which anyone can check, fishers may create false log books, hidden holds, or use illegal nets. When the Spanish fishing vessel *Estai* was captured by Canadian authorities in what came to be known as the "turbot war," they found that the majority of the turbot on board were too small to have been caught legally. The ship also had hidden holds, containing other species on which there was an international moratorium, and a net with illegal mesh size. Although Canada pushed the boundaries of international law in detaining the vessel, if it had not done so no one would have known the details of the Spanish vessel's illegal fishing behavior.

But because states are not obligated to join international agreements in the first place, they may be unwilling to create – or join – agreements with serious penalties for non-compliance, or with intrusive enforcement and monitoring procedures. Fishing regulations are more difficult to enforce than most international obligations, because so many different actors are involved in many actions. Even if a given fishing vessel follows the rules on one trip, there is nothing to guarantee that it will do so on its next voyage. And at any point in time there could be hundreds of thousands of vessels fishing worldwide in areas over which RFMOs have jurisdiction.

Most RFMOs rely on self-reported data, as do international agreements generally. States are obligated to report catches by vessels registered with them, and to ensure that their ships are following the rules. Vessels have an incentive to underreport catches and states have little incentive – and sometimes insufficient capacity – to police their own ships. In the face of likely non-compliance, some RFMOs have worked to implement additional monitoring procedures. The most labor-intensive of these are observer schemes, in which nationals from one state observe catches on vessels registered in another state. This approach was used in the International Whaling Commission in the 1970s at the point at which only a few states were still engaging in commercial whaling. Other fisheries commissions experimented with observer exchanges, including CCAMLR and a robust current one in the Inter-American Tropical Tuna Commission, ensuring that dolphins are not harmed in the course of tuna fishing. Observer schemes are expensive, however, and observers regularly report harassment or bribery in efforts to prevents them from reporting illegal behavior.[12] Catches can also be inspected at the point of landing, but it may be impossible to determine at that point whether the fish were caught using the proper procedures.

More recently RFMOs have been relying on technology as part of their monitoring operations. Satellite tracking equipment can determine where a fishing vessel has traveled, to ascertain whether it has fished in the authorized locations. Vessel monitoring schemes are now used in CCAMLR, the Northwest Atlantic Fisheries Commission, the Northeast Atlantic Fisheries Commission, as well as in national waters. This approach can address compliance with some aspects of fishing rules (where does a vessel catch fish?) but not others (how much fish does it catch and how does it catch them?). Compliance will continue to be a difficult aspect of international fishery management.

The difficulty of ensuring compliance with the rules makes it harder to agree on rules in the first place. With the existence of complete cooperation with well-crafted rules, everyone will benefit in the long-run, at the cost, perhaps, of shared short-term restrictions. But if states or fishers cannot be assured that other states or fishers will take on or uphold the rules, they face the possibility of losing twice: they may suffer in the short-term from the restrictions they undertake, but fail to benefit from the long-term protection of the fishery if others are failing to live up to their obligations. In a situation in which you are not certain that others will follow the rules, you are unlikely to be willing to agree to strict regulation. For this reason, measures that increase monitoring can increase the likelihood that states will take on stricter rules.

Other behavior changes occur as side-effects of the RFMO regulatory process. Because organizations focus on regulating catches of a particular species, these regulations can be ineffective at addressing bycatch. Many RFMOs have started addressing bycatch issues, often by mandating gear that is less likely to catch unwanted species, or by mandating observers to ensure that bycatch reduction efforts are followed.

The most common regulation internationally is catch limits. Because fishers are limited in the numbers (or weight) of fish they can catch, they seek to bring in the highest value fish possible, which will be the most lucrative. They may therefore engage in "high-grading," the process of catching more fish than they are allowed to land, keeping the highest value fish, and throwing the remainder overboard. Although such practices are clearly bad for the health of the fishery (since the fish returned to the ocean almost always die) and violate the spirit of the regulatory process, it is almost impossible to detect or prevent them.

A second type of effect occurs across fishing regions. Because RFMOs regulate by region or by species, their only jurisdiction is over member state vessels fishing in the regulatory region for the regulated species. But fishers lose income when they cannot fish. Some vessels, when faced with a seasonal closure or catch limits in one region will simply move to a different region or species and continue to catch fish legally. While this piecemeal approach might, at any one time, manage to protect the fish in a particular region, it does so by encouraging the fishing effort to move elsewhere on the ocean, keeping fishers in the business by exploiting stocks not yet protected.

Another important population is those fishers who remain outside of the regulatory process altogether. Because states are sovereign entities with no overarching authority (such as a world government) to require them to take on international rules, they must agree to join the relevant international regulatory bodies and cannot be made to do so without their consent. Some states have pursued a strategy of remaining outside of RFMOs so that their fishers can fish in international waters without restriction. Moreover, because states can choose to offer registration to vessels not owned by their citizens, if they wish to, some states have lured ship registrations intentionally by remaining outside of these international rules. This phenomenon of flags of convenience is addressed further in Chapter 3. States that offer such registration options to vessels owned by non-nationals can then earn additional state revenue from the taxes and fees charged to ships for registration, and those ships are willing to pay because of the advantages of being able to operate outside the international regulatory process.

Other Regulatory Options

Driven, in part, by the failures of international or standard domestic regulation to protect fish stocks, scholars and policymakers have begun to innovate different approaches to addressing fishery management. The most prominent of these work to change the issue structure of fisheries, either setting certain areas off-limits altogether, or privatizing some aspect of control over fish resources.

Marine Protected Areas (MPAs)
One way to protect fish is to protect a specific area of the ocean, restricting the types of activities that can be conducted in the area, often disallowing fishing or other resource extraction altogether. Protected areas in the ocean are referred to as marine protected areas (MPAs); the areas that disallow harvesting altogether are sometimes called marine reserves, to distinguish them from marine areas protected in other ways. The areas chosen as MPAs are often those where the ocean's resources are particularly vulnerable. These often include spawning or nursery areas for fish species, or fragile habitat areas (such as coral reefs). Because some fishery protection includes the idea of closed seasons or areas, it can be difficult to determine when the practice began, but the earliest major MPAs were established beginning in the 1970s.

These areas have had beneficial effects on protection of fish species that spawn or shelter in them. Studies have shown that some areas closed to fishing (as well as other MPAs) resulted in increased overall biomass in the area and recovery of some fish species.[13] In other areas, species of concern have been bigger inside protected areas than outside, another indication of success. MPAs can also improve biodiversity more generally in the region, which helps both commercial species and non-target species affected by fishing. They are not a panacea, however. If fishing pressure is not reduced globally (as it has not been), fishing effort will simply move to other areas or other species. Although the most vulnerable species may be better protected and juvenile fish may have a greater chance of reaching reproductive maturity because of these areas, such an approach is inevitably piecemeal and does not address the broader problem of too much fishing on the oceans. Protecting one species or area by moving fishing effort elsewhere simply shifts the burden to other species and areas.

It is also difficult to enforce fishing rules or prohibitions in these areas. Although most MPAs are located in national waters, many of these are nevertheless quite far from shore, and it can be difficult to police them to ensure that ships are not breaking the rules. Some non-compliance may even be unintentional, as it is difficult to mark the boundaries of a protected area. These difficulties can be even more pronounced when dealing with high-seas MPAs. Complete prohibitions on harvesting are easier to enforce than are rules about what can or cannot be harvested using which fishing methods; in the case of a ban, any ship seen fishing can be assumed to be breaking the rules. Technology discussed in Chapter 3 (such as satellite tracking of vessels) can assist in this kind of enforcement.

The ten biggest nationally created MPAs account for nearly 80 percent of the total nationally protected area globally. The oldest of these is a National Park in the waters around Greenland designated by Denmark in 1974; the largest are the Phoenix Islands MPA designated by the Republic of Kiribati in 2006 (enlarged in 2008 to 410.5 km^2), the Great Barrier Reef Marine Park, designated by Australia in 1979 (at 344.4 km^2) and the Papahānaumokuākea Marine National Monument off Hawaii, designated by the United States in 2000 (at 341.1 km^2).

Most MPAs, however, are still fairly small; the median area is smaller than 5 km^2, with at least 35 percent of all areas smaller than 3.5 km^2. Although they are usually set up to contain the entire habitat of some species of concern (at least at some phase of the fish's lifecycle) they can rarely fully contain the species they intend to protect. MPAs in the tropics, surrounding reefs, have had much better success at protecting fish, because these species spend most of their life in the protected area. But in colder areas or those places further out to sea, where most of the world's commercial fisheries are found, fish species range much further afield and are less likely to be protected through a significant portion of their range or lifecycle. Areas or species of high commercial fishing are less likely to be protected.

And creating international MPAs has proved difficult. Such areas could be created through cooperation among governments with contiguous EEZs, to protect areas that cross their marine borders, or through international organizations or regional fisheries management organizations establishing international areas. The parties to the Convention on Biological Diversity passed a decision in 2004 to encourage the creation of regional marine protected areas to conserve ocean biodiversity, with a target deadline of 2012. An early effort by the international organization OSPAR (comprising 15 countries off the west coast of Europe) to establish an MPA in the Mid-Atlantic Ridge in 2009, failed; the organization planned to consider this proposal again in late 2010, as well as nominate additional sites for international MPAs. The parties to the Antarctic Treaty (and the related Convention for the Conservation of Antarctic Marine Living Resources) did manage to establish an MPA in 2009 in the Southern Ocean off Antarctica, covering 90,000 km^2 around Orkney Island, after efforts to do so had failed for years. The establishment of this protected area is a major achievement, driven in part by the need to protect fish resources used by other endangered animals in Antarctica, such as penguins, albatrosses, and petrels.

Currently there are more than 5,000 MPAs, covering 2.85 million km^2, or just shy of one percent of the ocean's area.[14] Of all protected areas, only one percent are closed to resource harvesting altogether. Although some conservation organizations have proposed networks of MPAs that would cover up to 20 percent of the ocean's area, current measures fall far short of what would be needed for this policy approach to have a meaningful impact on global-scale fishery conservation.

Individual Transferable Quotas (ITQs)
A recent trend in environmental policymaking has been the use of market-based incentives to influence, rather than mandate, specific limits to behavior. Taxes or fees on pollution, for instance, force those who pollute to incur a cost for their behavior. This kind of policy instrument does not require anyone to change behavior; it just makes that behavior more costly. Removing subsidies has a similar effect, because subsidies keep a subsidized industry from paying the full cost of its activity. If someone else were not paying part of the costs, fishers would, in some circumstances, no longer find it profitable to catch fish. These market mechanisms allocate reductions in problematic behavior in the most efficient way. If you have to pay a tax on the amount of pollution you generate, you are likely to reduce your pollution if doing so costs less than paying the tax. If doing so costs more, you will instead make the rational decision to continue your behavior and pay the tax.

The biggest downside, however, is that these types of measures cannot enforce – because they are not designed to enforce – specific limits. If there were a tax on fish landings, it would raise the cost of fishing, but anyone willing to bear that cost would be able to fish any desired amount. Fishing behavior might change under the policy, but it would not be clear in advance exactly how it would change. In the face of severe overfishing, imposing a tax on fishing would be risky, and, perhaps for these reasons, has not been used in any widespread way in fishing regulation.

Another regulatory approach makes use of the incentive-based and efficiency benefits of market forces but combines them with the advantages of specific limits. This policy tool issues what are sometimes called tradable permits. (In the air pollution issue where this type of regulatory tool was first adopted, it often falls under the heading of cap-and-trade.)

The principle is that a limit on behavior is imposed, but then it is distributed as a property right that can be traded so that it can be implemented in the most economically efficient way through the mechanism of the market.

In the world of fisheries regulation, this type of regulation is generally called either Individual Transferable Quotas (ITQs), Individual Fishing Quotas (IFQs), or catch shares. Their initial implementation has happened almost entirely in the domestic context. The basic approach is that a specific limit is set on fish catches in a given season. That limit is then distributed (in a number of possible ways) across individuals or companies. Once you have a right to a proportion of the catch, you can then use it however you want. You can simply catch your allocation, and the advantage is that you do not need to worry about catching your allocation quickly before a seasonal closure is declared; you have an absolute individual right to that quota of fish. You may instead choose to sell all or part of your allocation to anyone who will buy it. Those who can fish at the lowest cost are likely to buy the allocations (and those for whom fishing would be more expensive can earn a profit by selling them), leading to an increase in fishing efficiency. (In practice most ITQ systems put some restrictions on sale, to preserve size diversity among fishers or to prevent industrial concentration.)

The first such system was introduced by New Zealand in the mid-1980s initially for the deep sea fish stocks in the country's EEZ and then eventually inshore stocks as well. This first experiment sold rights to fixed tonnages, but the system was not responsive to stock changes or incorrect estimates of sustainable yield, and in the 1990s the management system switched to the now-common share of whatever quota is set. Other early users of ITQ systems were Australia and Iceland. Canada and the United States have adopted this approach for managing some of their fish stocks as well, including a recent application to the groundfish fishery off New England.

Conceptually what an ITQ system does is undo the common-pool resource nature of a fishery by privatizing rights to fish. Because it does so by fishing vessel or fishing company it does not recreate the CPR problem at a different level the way that EEZ enclosure does. One of the main advantages of an ITQ system for fishing is that it decreases the "race to fish," and its associated problems, including overcapitalization. It is also likely to improve safety of fishing operations; fishers do not feel the need to go out in dangerous weather since they can catch their allocation whenever they choose. This approach does not, however, address potential problems of bycatch or high-grading, and, without monitoring, does not do away with the incentive to provide false information about catches. Depending on how the total allowable catch for a fishery is calculated this system may also not eliminate the political pressure to set total quotas higher than scientific sustainability estimates justify, because fishers rarely want to see their catch allocations decrease.

The effect on fish stocks is mostly positive. A study of fisheries from 1950 to 2003 examined 11,135 fisheries, 121 of which used some form of ITQ system, and found that ITQ-regulated fisheries were less than half as likely to collapse as those regulated by other measures, controlling for species and region. Many of these fisheries were, before adoption of ITQ systems, on a trend that suggested collapse, so it appears likely that the ITQ instrument itself accounts for the dramatic difference.[15]

An interesting question is whether such a system could work at the international level. Currently all ITQ systems are domestic, which has the advantage of bringing national-level enforcement (and universal participation by all who might engage in fishing in a region) to back up the property right aspect of such a system. There are academic discussions about whether and how such a system could work in an international context, such as the high seas tuna fisheries in various regions.[16]

Conclusion

As fishing intensity and technological sophistication increased, and stock declines became evident, governments attempted to protect fishery resources. Although early self-governance was possible by small-scale communities working to protect geographically bounded fish stocks, most fishing regulation now involves national or international rules. The scale of national regulation changed with the creation of 200-mile exclusive economic zones, within which countries were given the rights to manage marine resources. Although this approach might have been expected to remove the commons-like characteristics that make fishing regulation difficult, most countries instead chose to subsidize their domestic fishing industry and a national-level race for fish resulted. Countries that do not fully fish their own EEZs have often sold the rights to other countries to fish there, but have rarely been able to impose or enforce effective catch limits.

International efforts, focusing primarily on imposing catch quotas or gear restrictions, have also often failed to protect fisheries. The regional fisheries management organizations that oversee international fishing regulation are composed of states, many of which are reluctant to impose short-run costs on their fishers for long-run conservation benefits. Quotas are thus often set higher than scientific advice suggests, and ships and states find ways around even those rules to catch more fish than the stocks can support.

Some newer approaches show signs improving management. Marine protected areas can set parts of the ocean off-limits to certain activities or to harvesting altogether, and they can be especially important for protecting spawning areas or juvenile fish. Individual transferable quotas can decrease the incentive for overcapitalization by allocating individual (rather than collective) rights to fish. Although both these approaches show promise for improving fish stocks, the underlying problem of too many fishers catching too few ocean fish remains. The next chapter examines whether farming fish can address this underlying problem.

CHAPTER FIVE

Aquaculture

Aquaculture is the technical term for fish farming. It is not a new phenomenon – on the contrary, it has been around for millennia. But it has expanded rapidly over the past few decades, both in the volume of fish farmed, and in the breadth of species farmed. The proportion of the world's total seafood production accounted for by aquaculture increased from 13 percent to more than 35 percent in the period between 1990 and 2007.[1] With the production of the global capture fishery having peaked at a little over 90 million tonnes a year, all of the growth in global production of seafood comes from the rapidly growing aquaculture sector. To the extent that global capture fisheries are at or near (or even beyond) their maximum sustainable yield, future growth in the supply of fish and other seafood can be expected to come exclusively from aquaculture.

This sector is dominated by developing countries. China alone accounts for almost two-thirds of the global total aquaculture production as measured by weight (although under half by value), and the top five aquaculture producers globally – India, Vietnam, Indonesia, and Thailand follow China – are all developing countries (again by weight; measured by value Japan comes in fifth).[2] A large majority of the world's fish farming happens in Asia – almost nine-tenths by weight, and about three-quarters by value. The differences between weight and value are caused by the different kinds of aquaculture practiced in different places. Most of the fish farmed in Europe and North America are species that have a high price per weight, such as salmon and oysters, whereas much of the aquaculture in Asia is of lower-value species such as carp, grown with less capital for a more local market.

The range of activities undertaken under the heading of aquaculture, and the range of scales on which it is practiced, is as broad as for capture fisheries. The range extends from small ponds in which a few herbivorous fish, such as carp or tilapia, are raised for local consumption, to massive industrial operations, involving millions of fish, that raise open-water predator species in pens at sea. Many aquaculture practices, including the small-scale fish pond example, are essentially benign. But as the scale of fish farming operations increases, and as it comes to involve species that are higher up the food chain, potential problems associated with aquaculture multiply. Unfortunately, it is the species that have the greatest value on the global market, such as salmon, shrimp, and tuna, that generate the greatest environmental challenges. These challenges include pollution from high concentrations of animals in confined spaces, the spread of disease among confined and in-bred populations, ecosystem disruptions, and threats to the biological integrity of the species in the wild.

Some of the issues facing aquaculture are analogous to those in agriculture, particularly in animal husbandry. Both are activities that, when undertaken at the local level in traditional ways, can be environmentally benign. But in both cases, increases in scale, increases in levels of industrialization and mechanization, and increasing focus on large-scale monoculture production, generate both pollution problems and threats to the species being farmed. The problem of concentrated animal waste from confined animal feeding operations, or CAFOs, for cattle and pigs is mirrored by the output of large-scale shrimp farms, and the problem of increased susceptibility to disease in CAFOs, requiring the wide-spread prophylactic use of antibiotics, is also to be found in shrimp- and salmon-farming operations.

In some ways the political economy of aquaculture is similar to that of capture fisheries. Production ranges from the artisanal to the capital-intensive industrial, and the product enters the same global food supply chain. In other ways, however, the two approaches to producing fish are quite different. The greatest management challenge of capture fisheries, deciding on and enforcing quotas that prevent over-fishing, does not apply, since aquaculture does not face the same fixed limits to scale. And since aquaculture operations are geographically fixed, the international politics of fisheries management cooperation do not apply. In some ways aquaculture resembles agriculture more than it does capture fisheries. The greatest challenges created by increases in scale are posed by pollution and biological diversity problems rather than overcapacity, and the issue of genetic engineering that confronts international trade in key agricultural products is increasingly relevant to industrial aquaculture as well.

This chapter addresses all of these issues in turn. It first presents the history and development of aquaculture, and current patterns of industry production and growth. It then examines various forms of aquaculture, through specific examples of five farmed species: tilapia, oysters, shrimp, salmon, and tuna. It discusses both the potential advantages and the potential problems of each. Finally, the chapter explores the relationship between the aquaculture industry and traditional capture fisheries, and the extent to which the current rapid expansion of aquaculture globally can help to ameliorate pressure on global capture fisheries.

A Brief History of Aquaculture

Aquaculture has probably been practiced in some form since before there were written records. Scholars in China suggest that the practice began there nearly four thousand years ago, and others point to eel-raising in Australia two thousand years before that. The earliest written discussion of aquaculture comes from approximately 500 BCE in China, focusing on carp farming. There are also historical records suggesting that ancient Egyptians may have practiced aquaculture, and oysters were farmed during the Roman period.

Most early aquaculture was of freshwater species. Initially the practice may have sprung up opportunistically, when floods receded and left behind small bodies of water with juvenile fish, but expanded to include collection of immature fish to be raised intentionally in enclosed bodies of water. Early coastal aquaculture may have begun opportunistically as well, as tidal fluctuations left brackish ponds or swamps in which crustaceans or mollusks could easily be contained.

An essential stage was the development of processes to fertilize fish eggs separately from the natural reproductive cycle and raise the offspring that resulted; this practice may have originated in Europe in the 1700s. Despite the widespread adoption of aquaculture in traditional cultural practices and in many places around the world, it did not really take off on an industrial scale until the 1970s.

Contemporary Aquaculture

The aquaculture industry is in the process of expanding rapidly. Whereas the global output of capture fisheries has been static for several years, the global output of aquaculture been growing at a rate of about seven percent per year,[3] faster than the overall growth of the global economy, and much faster than the growth of global population. In fact, the growth has been sufficient to keep the amount of fish available per person growing in the face of global population growth despite the flattening and even slight decline of capture fishery production. Most of this growth has been in developing countries, led by China. Much of the aquaculture in developing countries is of relatively low-value species, and is farmed for local consumption. Aquaculture in developed countries is more likely to be of high-value species, such as salmon, although many high-value fish, notably salmon and shrimp, are also farmed in developing countries.

As is the case throughout this volume, the term "fish" as used here refers to seafood in general, rather than fish specifically. Of the 50 million tonnes of seafood produced annually in aquaculture, just under two-thirds are vertebrates (what we normally consider to be fish), about one-quarter are mollusks (mostly bivalves such as oysters and clams), and most of the rest are crustaceans (mostly shrimp).[4] In addition, about 14 million tonnes per year of marine plants (mostly various kinds of seaweed) are grown, but, while these count as a form of aquaculture, they are not discussed here. The most widely farmed kind of fish globally is carp. Of the ten most farmed species by weight, six are varieties of carp, accounting for more than one quarter of the overall total. These also account for almost one fifth of the global total by value, about $17 billion – carp are relatively inexpensive, but the volume grown adds up.[5] Other major kind of fish by value, with over $10 billion per year farmed, include shrimp, salmonids (salmon and trout), and bivalves (primarily clams, oysters, and scallops).

Distribution of farmed fish is generally not distinct from the distribution of caught fish. Aquaculture production for the global market is more likely than capture fishery production to be distributed through corporate supply chains, however, and less likely to be distributed through auction markets.

Similarly, aquaculture production for the global market is less likely to happen at artisanal scales, and more likely to happen at industrial scales, than is the case for capture fisheries. Industrial-scale aquaculture can produce fish more reliably, and with less variance amongst individual fish within a species, than capture fisheries can. These fish are therefore well suited as inputs into the industrial food supply and processing chain. Artisanal aquaculture production, common in developing countries and rarer but existing in developed counties, is likely to either be consumed onsite or to be sold in local markets.

Aquaculture Examples

Aquaculture can happen on a wide variety of scales, using levels of technology ranging from the preindustrial to the computerized. It can also be done in a wide variety of settings, from natural or artificial inland ponds, to lakes, streams, and rivers, to tidal flats, to pens in the open ocean. And it is involved in growing the full range of commonly eaten seafood species, from simple bivalves to the almost warm-blooded bluefin tuna. Five particular examples of forms of aquaculture serve to illustrate this range of activities, and both their potential and problems. The species examined are tilapia, oysters, shrimp, salmon, and bluefin tuna.

Tilapia
Tilapia is a useful first example because it represents both a historical form of and the largest single contemporary category of aquaculture. Tilapia is an inland, freshwater, omnivorous fish, a category including the most common single kind of farmed fish, carp (both tilapia and carp are categories that include several related species). Tilapia is discussed as an example here rather than carp because tilapia is

likely to be more familiar to contemporary western consumers. It is a tropical fish originally found (and fished) mostly in Africa and the Middle East, but can now be found in tropical and subtropical climates globally. Tilapia will eat anything, but generally prefers vegetation and other aquatic detritus. This omnivorousness makes the species both easy to establish in new habitats, and useful for cleaning up freshwater sites of excess plant and other undesirable materials (such as fecal matter from other fish, which tilapia can consume in great volume). As such, tilapia can be used to purify and control vegetation in water even when not being farmed. The fish are easy to farm, both because they eat almost anything and because they have a simple reproductive cycle, and grow particularly fast.

Farming tilapia can be relatively environmentally benign, but can also be environmentally problematic, depending on how it is done. Some of the potential negative environmental effects involve pollution, and are generally quite localized. The environmental impact of tilapia farming depends primarily on how intensive it is, meaning how densely the fish are packed together. If they are raised at a concentration at which they can thrive on the locally available food supply, tilapia aquaculture can be environmentally beneficial. For example, tilapia can be raised in paddies alongside rice cultivation. The fish can actually improve the yield of the rice crop while at the same time decreasing chemical use, farming effort required, and insect problems. The fish eat excess underwater foliage, insects, and other detritus, and in turn help to fertilize the crop.

Tilapia can only be farmed in this way at low densities, however, because the amount of food naturally available in rice paddies, and the ability of plants to absorb nutrients excreted by the fish, is limited. Closed-system aquaculture, that requires no outside inputs and is therefore environmentally sustainable, generates a relatively low yield for the amount of effort put in and the amount of land used. Raising tilapia and other similar omnivorous freshwater fish at an industrial scale requires much greater concentration, meaning that many more fish are kept in a similar volume of water. The greater the concentration of fish, the greater the potential environmental effects. These effects occur both in inputs and outputs from the aquaculture process.

A key input is food. In the rice paddy aquaculture process, no outside food is required, because the tilapia can feed themselves from plant and insect matter. But when the fish are concentrated to the point where the available food supply is insufficient, or when the fish are grown in an enclosed setting such as tanks or artificial ponds in which natural food is not available, they must be fed food from outside their immediate environment. In the case of omnivorous fish such as tilapia or carp, this food can be plant matter or animal protein. In the former case, aquaculture is producing fish protein from plant matter, and therefore probably taking some strain off of global capture fisheries. But tilapia are sometimes fed fish meal, both because it is cheap and because a more protein-intensive diet makes them grow faster. When fed fish meal, farmed tilapia are actually increasing the strain on the supply of fish in the wild, because over the course of their lives they can eat more than their weight in fish meal.

The key output is fecal matter and other fish residue, as well as residues from any chemicals used in the farming process. In a rice paddy, the residual matter is absorbed by plant life, and therefore creates no negative environmental impact. But the more concentrated the fish, the more concentrated the waste, and when fish are raised in artificial tanks or ponds without plant life, there is no natural ability to absorb residual waste at all. In tanks the water can be cleaned before it re-enters the general water system, although the waste still needs to be disposed of in some way. When concentrated aquaculture happens in natural bodies of water, however, the waste matter is unlikely to be contained effectively, and generally ends up polluting the rivers or lakes that the water flows to. Tilapia aquaculture in Sampaloc lake in the Philippines, for example, led to a deterioration in lake water quality once the Tilapia reached a concentration that required them to be fed by external sources of food.

Farming fish like tilapia is thus in important ways like farming land animals. In both cases, raising the animals as part of an integrated farming system, in which plant waste feeds animals and animal waste in turn feeds plants, can be largely environmentally benign. It can result in the use of fewer external inputs, and create less pollution, than farming either plants or animals alone. But when the process is industrialized, and when there is a focus on monocultures (the raising of single species) at the expense of integrated agriculture or aquaculture, both inputs (food and chemicals) and outputs (pollution from waste products and chemical runoff) increase. The amount of input and output per animal increases with the level of industrialization. In this sense, a big tilapia farm might be like a large livestock feedlot, with animals fed on imported corn and producing huge lakes of feces. Few tilapia farms approach this scale, but the danger is real. Tilapia farming can be relatively environmentally benign, but we cannot safely assume that it will always be so.

Another similarity between aquaculture and agriculture is that intensive farming can yield food that is less healthy than fish or crops grown in a less intensive, more sustainable manner. As diet or growing conditions move farther away from what they would be like in a natural habitat, the resulting fish or crops can become less nutritious. An example of this phenomenon in the case of tilapia is omega-3 fats. These have been discovered recently to be important nutrients, necessary for, among other things, effective brain functioning.

Omega-3s are found in high concentrations in seaweed and other marine plants, and are therefore often found in high concentrations in fish as well, particularly fish like tilapia that tend to eat mostly marine plants. But intensely farmed tilapia may well not be eating marine plants at all – they may be eating whatever feed is cheapest at the time. The fish will still produce omega-3 fats, but the wrong kind of omega-3s (short-chain fats, rather than the long-chain fats that are best for us). So eating fish farmed at industrial scales may not be as good for us as we think. This issue is relevant to many other farmed fish as well, including salmon.

The final key danger to the farming of species like tilapia discussed in this section is the threat to biodiversity posed by the globalization of aquaculture. Although species of tilapia are native to Africa and the Middle East, the fish can now be found throughout the tropical and subtropical world, and even in pockets of warm water in some temperate climates. The reason that tilapia are so useful for aquaculture, and are imported for aquaculture operations globally, is that they are hardy. They can eat anything, and survive in a wide variety of specific habitats, as long as they are warm enough. This same ability to thrive in a variety of habitats, however, makes them a particular danger as invasive species.

And they are indeed an environmentally problematic invasive species on a global scale, displacing native species of fish from ecosystems around the tropical world. For example, tilapia were introduced to the Galapagos Islands, where geographic isolation has resulted in the evolution of species that are ecologically sensitive; they threatened to out-compete endemic species until eventually eradicated by a project funded by international development and aid agencies.

Nor are tilapia unique in this respect among freshwater farmed species. Carp can pose an even greater danger as an invasive species, because the species' greater tolerance for cold water increases the range of habitats that it can invade. For example, Asian carp have become established in much of the Mississippi river basin, and are currently threatening to invade the Great Lakes via the Chicago Sanitary and Ship Canal connecting the Mississippi system with Lake Michigan. Once the species has invaded Lake Michigan it would be impossible to keep carp from migrating to the rest of the Great Lakes. Asian carp reproduce quickly and grow to sizes of up to 100 pounds, eating 40 percent of their bodyweight daily. They can cause devastating changes to ecosystems they invade, eating food that would have supported other species. In some bodies of water they have invaded in the United States they now account for more than 95 percent of the biomass.[6] Several organizations, including the federal and state or provincial governments in both the United States and Canada, as well as industry representatives and non-governmental organizations, have formed the Asian Carp Regional Coordinating Committee to try to keep the carp out of the Great Lakes. The United States Environmental Protection Agency alone has budgeted $13 million for this effort.

The process of domesticating tilapia for aquaculture is also affecting the genetic diversity of the species. The "natural" (or pre-human) versions of tilapia species are rapidly disappearing – those found in the wild in the contemporary world are more likely to be crossbred than natural species. This process can undermine genetic diversity within species, meaning that diseases or pests that would previously only have affected a subset of tilapia varieties are now more likely to affect a broader range of them. This problem happens not only with many kinds of aquaculture (it will come up again in the discussion of salmon, below), but also with agricultural monocultures.

Oysters

Oyster aquaculture is different from farming most kinds of fish because the oysters more or less stay in the same place. There is therefore no need to build ponds or tanks, or to install nets or pens, to contain them (although some forms of oyster farming do involve tanks, hatching oysters in onshore facilities and placing them in beds when they are old enough to attach themselves). An oyster farm in an area that is a natural habitat for oysters, such as a river estuary, can thus be ecologically not that different from naturally occurring oyster beds. Oysters (and other bivalves) in their natural habitat fulfill the useful ecological function of reducing the amount of nitrogen in the water. Reducing nitrogen in turn reduces the amount of phytoplankton in places where it has grown to excess, resulting in cleaner tidal shallows that can support a greater variety of marine life. Since farmed oysters get their food from the surrounding habitat, rather than being fed industrially, many of the problems of excessive concentration and crowding found in industrial aquaculture are not a problem with oyster farming. Because they rely on a natural food supply, the oysters cannot be concentrated beyond the carrying capacity of their environment. (It is nevertheless the case that oyster aquaculture removes sources of nutrients from the aquatic environment, because after they have grown to harvest size the animals themselves are removed.)

Because oysters can play a useful role in an ecosystem, oyster aquaculture can provide environmental services. An example of a place where these benefits have been realized is the Chesapeake Bay on the east coast of the United States. In this area the abundant natural oyster population was largely fished out over the course of the twentieth century, depriving the ecosystem of a natural filter and resulting in both the economic loss of a fishery resource and a dirtier estuary. Government agencies and environmental non-governmental

organizations are now reseeding the oyster stock throughout much of the Bay. This restocking has the twin effects of
reviving the fishery and helping to restore the Bay's marine
ecosystem.[7] There is a fine line here between oyster aquaculture and reseeding a natural breeding stock. But the fact
that it is a fine line suggests that it is a relatively ecologically
benign activity. The lack of environmental threat, however,
assumes that the reseeding is done with native species of
oyster. Seeding with non-native species risks the most ecologically harmful of effects of oyster farming (and other forms of
bivalve aquaculture), the introduction of invasive species.

Bivalves would not at first glance seem a major danger
as invasive species, because they stay in one place. But
because they stay in one place, their mating patterns involve
spreading eggs and sperm as far as possible throughout the
available current (although many swimming fish do this as
well). Ironically, the fact that they are stationary creatures
in a way increases the danger of invasion from aquaculture.
Since they cannot swim away, they do not need to be netted or
caged (although they are sometimes grown in bags). There is
therefore often no barrier preventing them from reproducing
beyond the farming environment, in the broader ecosystem.
Aquaculture is not the only, and probably not even the most
important, source of invasive bivalves. One of the most problematic invasive bivalves, zebra mussels, spreads through
shipping (in ballast water and on anchor chains) rather than
farming. These invasive mussels clog water intake valves, and
cause damage to ships, harbors, and power and water treatment plants. When oyster or clam farming uses non-native
species (generally because they are more lucrative than native
species) it can introduce species that then compete directly
with (and sometimes out-compete) native species. The Pacific
oyster, a species native to East Asia, has been introduced as
a farmed species in many areas, in some of which, such as

Washington State in the United States and the Wadden Sea area on the coasts of Germany and the Netherlands, it is displaying invasive characteristics. Other species that can travel on the shells of oysters can also become invasive as a result of oyster aquaculture.[8]

The problem of invasive species in oyster farming is not insurmountable. Farming local species eliminates it. And recent trends in oyster aquaculture involve mating different kinds of oysters. The resulting oysters are often bigger than their parents, and are unable to breed themselves. This approach creates a situation analogous to agricultural crops that cannot reproduce, such as hybrids or many genetically modified cereals. It complicates the process of farming, and makes the whole process more capital intensive. But it does solve the problem of invasive species – oysters cannot be invasive if they cannot breed.

Shrimp

Shrimp are among the most valuable products of aquaculture. Pacific white shrimp, in fact, while only the world's sixth-biggest aquaculture product by weight, has been its single most valuable species since 2004, yielding almost $9 billion in 2007. Tiger prawns, the next most important shrimp-like species, were worth an additional $3 billion that same year.[9] Shrimp are worth discussing as an example of carnivorous fish because this category of aquaculture necessarily increases, rather than ameliorates, pressure on the world's capture fisheries.

A large majority of shrimp farming is done in the developing world. It is generally done at an industrial scale, but in a large number of relatively small operations. Furthermore, it is often done in countries with inadequate regulatory structures, particularly in south and southeast Asia. Since environmental and health-related practices vary both across countries and across operations within countries, it can be difficult to track the provenance, and therefore the environmental impact, of particular shrimp.

Shrimp aquaculture creates problems of habitat loss, waste, and disease, and also depletes the world's capture fishery. Habitat loss refers to the destruction of natural habitats, either of the species being farmed or of other species, in the process of creating and running fish farms. Habitat loss can be a problem with any kind of aquaculture, but the problem is particularly acute with shrimp. Shrimp farming is often done in coastal shallow water in the tropics. The construction of shrimp farming operations often displaces either mangrove forests, which in turn are key to the maintenance of functioning coastal wetlands, or beach habitats in which sea turtles had previously laid their eggs. It is because of this loss of habitat, along with the effects of trawling for shrimp in the capture fishery, that some environmental organizations, such as Greenpeace, now recommend that ecologically conscious consumers avoid shrimp in general.

Shrimp farming also creates waste. As discussed with respect to tilapia, animal waste becomes more of a problem as the scale and concentration of an aquaculture operation increases. Because shrimp aquaculture is generally practiced on an industrial scale, it also generally involves issues of concentrated waste. And since the aquaculture itself is often conducted in the marine environment, the waste can be difficult to contain. Even when shrimp are grown to maturity in fully enclosed ponds that are separate from the shoreline, the waste needs to be deposited somewhere. There are shrimp farming operations that do contain their waste, and prevent contamination of surrounding waters and soils, fairly effectively. But since a majority of shrimp aquaculture occurs in countries in which environmental regulations are inadequately enforced, problems of contamination are fairly common.

Keeping animals in closer proximity to each other than they would be in the wild not only creates problems with concentrated waste, but creates health problems for the animals as well. This problem increasingly affects industrial animal farming on land, and is a problem with many forms of aquaculture. Confinement and close proximity can make fish more susceptible to disease, by increasing their stress levels and thereby decreasing the effectiveness of their immune systems. At the same time, it increases their exposure to both disease and parasites, by keeping them in close proximity to many other fish. If one shrimp in a shrimp farm becomes infected, it is likely to pass the disease on to more shrimp than would be the case in the wild, simply because it comes into contact or close proximity with more of them. The larger the scale of the operation, and the smaller the volume of water each individual is allowed, the greater the potential problem. Concentrated shrimp farming can lead to large-scale die-offs within aquaculture operations, and the transmission of disease across operations.

The standard solution to the problem of animal health in industrial-scale farming on land is antibiotics, and this solution is increasingly being used in shrimp aquaculture. To pre-empt the spread of infectious diseases, shrimp are given antibiotics prophylactically, mixed in with their food. This practice would be the equivalent of feeding people antibiotics on a daily basis to prevent illness. The antibiotics do help prevent the spread of infectious disease. But their use in this manner has negative side effects. Widespread and frequent use of antibiotics encourages the evolution of antibiotic-resistant organisms, which makes the potential for disease, not only in aquaculture, but also in the general population, greater. Concentrations of antibiotics also remain in the farmed fish, which may have health effects in humans (or may not – the evidence is not yet clear on this issue). The use of antibiotics is not specific to shrimp aquaculture; it can be found in some tilapia farming operations, and can be a problem in salmon farms as well, as discussed below. The denser the concentration of fish in any aquaculture operation, the more likely that widespread antibiotic use will be necessary to keep the fish alive.

Although many of the issues discussed in this section are common to aquaculture (and in some cases to industrialized animal farming in general), there is an additional problem particular to the farming of carnivorous species, such as shrimp (and, as discussed below, salmon and tuna). Growing a carnivorous species for food requires that it be fed more animal protein than it produces. As a practice, therefore, it is necessarily a net destroyer of food. The same is true to a certain extent of raising any animals that are given feed, rather than being allowed to forage. Cattle, for example, consume ten times as many calories of grain as they produce in meat. But the case of shrimp farming is even more extreme, because the shrimp are, for the most part, fed fish from the global capture fishery. Since the shrimp are worth so much more by weight than the undifferentiated fish that they are fed, doing so makes economic sense. But, in a world in which fishery resources are under severe strain and the yield of the world's capture fisheries seems to have peaked, it patently does not make ecological sense. (It also only makes economic sense because the fish caught in the capture fishery are essentially "free" to those who catch them.) In other words, far from easing pressure on the world's capture fisheries, shrimp farming actually increases the pressure.

This problem can be ameliorated to a certain extent by increasing the ratio of vegetable matter in the food that the shrimp are fed. But this strategy does not get at the core of the problem. Even if shrimp or other carnivorous fish are fed a higher proportion of vegetable matter in their feed, they remain fundamentally carnivorous fish. The farming of carnivorous fish necessarily increases the pressure on the world's living marine resources, and at the same time decreases the total amount of fish available to people as food. This problem gets worse the farther up the food chain one goes, because the fish in question become less efficient at turning food into body mass. Thus the ratio of other seafood eaten to meat produced is worse for salmon than for shrimp, and worse for tuna than for salmon.

Salmon

Salmon are the second most valuable aquaculture species, just behind Pacific white shrimp. They are also the aquaculture product most likely to be familiar to the average western, and particularly North American, consumer. Most of the salmon consumed globally, and almost all of the Atlantic salmon, is farmed (six species of Pacific salmon are still fished, but the commercial Atlantic salmon capture fishery was almost entirely fished out by 1990). Norway is the largest producer, and until its recent difficulties with a disease affecting salmon stocks (described below), Chile was a close second, with Canada and the United Kingdom far behind in third and fourth place.[10]

Many of the environmental issues confronting aquaculture in general are problems for salmon farming, and are often necessarily worse for salmon than for many other kinds of fish. As carnivorous fish farther up the food chain than shrimp, salmon present an even more inefficient use of the world's marine resources. The other complicating factor is that salmon are anadromous, meaning that they spawn in fresh water but mature in the open ocean. Farming them therefore requires more infrastructure, and a greater capital investment, than is true of aquaculture for more sedentary species. Their anadromous nature means that farming salmon requires creating two enclosed environments for them, one mimicking spawn-ing rivers and one mimicking (and often existing in) the ocean. These two parts of salmon farming can be done as separate operations, but even so, significant capital is required for either part of the process, and capital outlay is required to purchase salmon fry (baby salmon) for salt water operations. This neces-sary capital intensiveness means that the sort of small scale, ecologically sensitive operations that are feasible in some set-tings for more sedentary fish are not economically feasible for salmon farming. Also, the natural setting for mature salmon is the open ocean, which cannot be replicated in aquaculture. Salmon farming cannot be practiced in a way that allows the fish to feed in a relatively natural way, because they cannot do that while simultaneously being confined.

Salmon aquaculture is therefore necessarily industrial, and generally practiced on a fairly large scale. For example, in 2002 the average salmon-farming company in Canada's province of British Columbia produced over 5,000 tonnes of fish.[11] All of the problems that arise with industrial-scale aquaculture, such as pollution from fish waste, and health problems caused to the fish by intensive monoculture cultiva-tion, thus plague salmon farming. Two particular problems of these types affect salmon aquaculture. The first is an infec-tious disease called infectious salmon anemia (ISA). ISA has devastated Chile's salmon aquaculture industry, which as recently as 2008 was one of the world's largest. Chile's output has fallen by over three-quarters, as ISA killed off the salmon themselves, and made the fish that survived less appealing to foreign markets. Both the economic and environmental effects of ISA on Chile and the Chilean aquaculture industry have been huge.[12] The extent of the damage from ISA, and the speed with which it spread, suggests that the health issues associated with aquaculture, particularly salmon farming, are quite real and can be devastating.

The main risk to salmon aquaculture is a parasite, sea lice. Sea lice attach themselves to and feed off salmon, and find salmon farms to be a congenial breeding environment. Mature salmon can survive moderate infestations, but can succumb if enough sea lice attach themselves. Juvenile salmon are much more sensitive, and may be killed by only one louse. Aquaculture operations can usually minimize the damage to their own fish through activities including chemical treatments (which themselves create ancillary environmental problems of chemical runoff). But when salmon farms are located (as most are) in environments through which wild salmon swim, the farms act as breeding grounds for sea lice to infect the wild stock. The effects of the resultant increase in infestation in the wild stock are unclear at this point, but might be devastating – there is some evidence at this point that sea lice are the cause of the recent rapid decline of salmon runs on the Canadian and American Pacific coasts.[13]

As is the case with shrimp, farmed salmon are often fed antibiotics prophylactically. Two issues arise with this practice specifically in the context of salmon. The first is the question of what antibiotics are being used. Some Chilean salmon farms were recently discovered to be using antibiotics that were not approved for use in animals destined for the United States market,[14] suggesting problems with antibiotic use in a globalized food production chain. The second is that over time habitual antibiotic use leads to the evolution of salmon that have weaker natural immune systems, because fish with weaker disease resistance that would otherwise die before breeding are kept alive by antibiotics, and are able to breed and pass on their genes. Were it the case that farmed salmon were kept entirely separate from the natural breeding stock, this problem would only affect fish farmers. But farmed salmon are kept in pens in the ocean and some inevitably escape. The widespread use of prophylactic antibiotics in farmed salmon thus weakens the disease resistance of the natural stock.

The inability to keep farmed salmon strictly separate from their cousins in the wild presents a real problem for salmon aquaculture, much as is the case for tilapia. But the problem with salmon is worse, because salmon must be kept in pens at sea, and therefore cannot be completely segregated from the natural stock. Farmed salmon present two distinct dangers to the natural stock. The first is as a source of disease, as noted above. The second is as a threat to the genetic stock of salmon in the wild. This genetic threat is twofold, because of species distinctions in salmon. Atlantic and Pacific salmon are genetically distinct, with one predominant species of Atlantic salmon, and six of Pacific salmon. In the Atlantic, since farmed salmon are not exposed to the evolutionary forces of their natural habitat, when they escape captivity and interbreed with the natural stock they weaken the survival traits of that stock. In the Pacific, there is no evidence that escaped Atlantic salmon can interbreed with genetically distinct Pacific species, such as Sockeye or Coho. But they do compete with wild stocks for reproductive resources, including both food and the best nesting spots.

A related issue is the possibility of genetic modification of farmed fish. While none of the salmon that makes it to stores or restaurants at the time of writing is transgenic (genetically modified), the science for genetically modifying fish is well developed. From an industrial perspective, this approach is also promising, with the possibility of creating fish that grow faster and are more resistant both to disease and to pests like sea lice. There is an enormous literature on, and debate about, genetic modification of food in general, that is beyond the scope of discussion in the limited space available here. With respect to salmon specifically, genetic modification introduces both potential environmental benefits and hazards. The benefits are fish that require less food (because they grow faster), important when the fish eat other seafood that needs to be caught; and aquaculture that requires fewer chemical and pharmaceutical inputs (if the fish can be created to have stronger resistance). The biggest hazard is that transgenic fish will escape and either interbreed with or compete with natural salmon, and thereby threaten the viability of the natural genetic code. Genetic modification of salmon is more advanced than with most other farmed fish, and is still nowhere near as common as with many food crops. But the issue is on the horizon. The US Food And Drug Administration is at the time of writing considering an application to approve for human consumption transgenic salmon that grows twice as fast as natural ones. The issue will inevitably attract more attention as the science progresses.

Bluefin Tuna
Bluefin tuna are also a carnivorous species. And they have a faster metabolism than salmon, meaning that they yield even less fish by weight per input of feed. It takes approximately fifteen pounds of feed to yield one pound of bluefin tuna.[15] The objection in principle to farming bluefin tuna is even greater than it is for salmon. But farming bluefin tuna is not an equivalent process to farming salmon and, as currently practiced, it is much more problematic. Bluefin tuna aquaculture has sometimes resulted in what people refer to as tuna laundering, the fishing equivalent of money laundering in finance.

To compare bluefin tuna farming to the farming of other fish is misleading, and bluefin aquaculture is often referred to as ranching rather than farming to highlight the distinction. With most other forms of aquaculture, the fish involved are bred in captivity. But bluefin tuna (and many other kinds of tuna as well) will not generally breed in captivity. Ranched bluefin are caught in the wild and fattened up in big holding pens at sea. If the fish that are caught for tuna ranching are fully included in quotas on bluefin tuna under international management, and the fish that they are fed in captivity are similarly caught within functioning sustainable management schemes, bluefin aquaculture need not be a direct threat to the management of the relevant capture fisheries.

The problem that arises, however, is that the process makes the monitoring and enforcement of quotas more difficult, both with respect to the bluefin tuna themselves and to the species that the ranched tuna are fed. Catch quotas are self-reported by those who catch the fish, and it is unclear at what point that reporting happens. But fish that end up in the bluefin aquaculture process, be they the tuna themselves or what they eat, may be able to avoid that process altogether.

The first issue is how, or whether, the bluefin tuna themselves are caught within the catch quotas of existing regulatory procesess. Some countries that ranch tuna in the Atlantic are not members of the relevant international management organization, the International Convention for the Conservation of Atlantic Tunas (ICCAT), and therefore are not bound by any catch limits that organization places on its member states. (The same could, in principle, be true of tuna ranching in other regions.) Even for those states that are ICCAT members, the organization's processes are not set up to oversee ranching operations, and the young fish caught to be taken to ranching pens may not be reported as tuna catches by ICCAT member states. Juvenile fish may also be caught for ranching operations before they have reached reproductive maturity and therefore never get the chance to reproduce, additionally decreasing the overall tuna stock. Systems to track individual fish from the capture fishery through the ranching process to market are inadequate, which also has implications for scientific research, since good statistics on size and location of tuna catches are not reported from tuna

ranching operations. Even if tuna caught for ranching operations are counted to be within catch quotas, quotas generally
work by weight, and the fish caught for ranching operations
are young and small. They would therefore be counted as a
small percentage of a country's bluefin tuna quota. And young
fish caught that die either on their way to the ranching pens or
before they are taken to market are not counted at all.

Tuna ranching can also be used as a way to reprocess illegally caught tuna. Under efforts to prevent fishing outside
of international regulatory processes (described in Chapter
4), ICCAT now requires all member states to land only fish
that can be documented as having been caught under ICCAT
rules. But to the extent that tuna ranching operations operate outside the ICCAT process, illegally caught wild fish
could be repurposed as ranched tuna, and thereby gain access
to key markets (primarily Japan, the United States, and the
European Union). It is this process that is sometimes referred
to as tuna laundering.

The second problem for global fishing regulation of tuna
ranching is whether the fish that are fed to the bluefin tuna
as they are being ranched is caught within any regulatory
process. The answer is probably no. The fish can be delivered
directly to the aquaculture operations, which are at sea, and
never be subject to the quota process. In the case of the prey
species the evidence literally disappears – it gets eaten by the
tuna. And tuna will not eat the sort of generic fish meal that
salmon will eat. They require fresh fish, putting additional
pressure on the capture fisheries of preferred prey species, in
competition with the needs of tuna and other predator species
in the wild.

Efforts have been underway for some time to figure out
how to get bluefin tuna to breed in captivity. Scientists have
now been able reliably to replicate the entire breeding cycle
in captivity through the use of hormone treatments. This

development has resulted in some commercial farming of captivity-bred tuna, although the ranching approach to bluefin aquaculture still predominates. Were the captivity-bred model to displace the ranching model, tuna farming would still face all of the problems of carnivorous fish farming, but would at least not be adding additional pressure to the bluefin tuna capture fishery. Scientists and activists are of mixed opinion on the efficacy of captive breeding for bluefin conservation. Some see it as taking pressure off the natural stock, and providing the potential for restocking breeding populations in the wild. Others see some of the problems with aquaculture in general, such as dilution of genetic stock in the wild and incubation of disease (and the inefficiency of catching fish to feed to tuna), as outweighing the gains.

Conclusions

Aquaculture has been practiced for millennia, but it is only quite recently that it has been practiced on a scale comparable globally with capture fisheries. At current rates of growth, it is entirely possible a majority of seafood consumed directly by people will be farmed rather than caught in a decade or two. Our collective consciousness of the role of aquaculture, and the need to pay serious attention to it, lags behind the reality of its explosive growth. From the perspective of developed countries the most commonly consumed farmed fish are shrimp and salmon. From a global perspective freshwater fish farmed in China are even more common. The breadth of industrial scales on which aquaculture is practiced is every bit as great as is the case for capture fisheries, as is the range of relationships with the global market. The number of farmed species is smaller than the number of species captured in the wild, but aquaculture is increasing not only in volume but in types of species, and new species are being created for the practice, both through breeding and through genetic modification.

The key question about aquaculture in a book on the geopolitics of fishing is what the effects of the practice are on the world's capture fisheries. Can fish farming alleviate some of the pressure on the world's natural fish resources? The five cases outlined in this chapter suggest that the answer is mixed, and is complicated. Aquaculture can certainly produce large volumes of food, and add meaningfully to the global supply of animal protein available for people to eat. Although farmed fish is often not quite as nutritious as caught fish (because it is less likely to have micronutrients such as long-chain omega-3 fats), it is more nutritious than eating no fish.

But does this extra fish decrease pressure on global capture fisheries? It depends on several factors, and on the specific kind of aquaculture practiced. With the farming of carnivorous species, the answer is likely no, because these species eat more fish protein than they produce. This fundamental logic cannot be overcome by better aquaculture practices. The farming of carnivorous species in this sense represents a net loss from the perspectives of ecology, marine conservation, and global fisheries management, in addition to any other problems that it may generate.

With the farming of species that can be fed vegetarian diets or that do not need to be fed, the ecological calculus is more likely to be positive. With many species aquaculture can be practiced in an ecologically benign, and in some cases even ecologically beneficial, way. But there is no guarantee that it will be. Even species that can be farmed in a way that makes environmental sense, *can* be farmed in an ecologically harmful way. Problems from aquaculture practices can include pollution from concentrated animal wastes, threats to the health both of the farmed fish themselves and their cousins in the wild, habitat destruction, and non-native species

invasion. As a general rule, the larger the scale of the aquaculture operation, the greater the threat of ecological harm,
because most of these problems are problems with the industrialization of food production, much as is the case with many
forms of agriculture. But this rule is not universal – an invasive species, for example, can be introduced by a small-scale
operator.

Large-scale operations are also most likely to be able to
afford experimentation, careful management, and the introduction of new techniques to manage these problems. And
many large aquaculture operations are indeed working to
decrease their negative ecological impacts. In part industrial
aquaculture firms are doing so because, as in the Chilean
salmon case, bad management can lead directly to a die-off of
the farmed stock. In part they are doing so out of direct consumer pressure, and out of fear of, or in response to, more
stringent government regulation. But consumer pressure is a
much more organized phenomenon in rich than poor countries, and richer countries are similarly able to regulate more
effectively than poorer ones. Consumer and regulatory pressure are therefore able to affect management standards in the
sorts of high-value species that are most likely to be farmed
in or exported to rich countries. In the developing countries where the majority of aquaculture is practiced, and is
undertaken for local consumption, the pressures to improve
management standards is likely to be weaker.

More broadly, aquaculture can indeed take some pressure off the world's capture fisheries, at some ecological
cost. But it is not, as some advocates claim, a panacea. At
moderate scales it can be ecologically benign, but the bigger
the scale, the bigger the ecological threat it presents. To the
extent that it displaces caught fish in particular markets, it
can help to preserve natural stocks. But to the extent that it
helps to create new markets for seafood, it can actually help

to undermine efforts to manage the world's marine ecosystems, even aside from the farming of carnivorous fish.
It is, in this sense, at best a useful pressure valve, not a
solution.

CHAPTER SIX

Consumers and Catches

To recap the core arguments of this book, both the geopolitics and geoeconomics of capture fisheries are fundamentally different from those of other resources. There are a few key sources of difference. One is the nature of the resource: there is a fixed upper limit on how much of the resource we can extract in any given year without the supply of the resource crashing. With almost no other resource is it the case that overexploitation in one year can lead so directly to disappearance in the next. This phenomenon is exacerbated by uncertainty: we often do not know exactly where the upper limit is, meaning that we must either underexploit as a matter of course or risk environmental disaster when we overestimate the carrying capacity of a stock (as we do on a regular basis). Even when stock collapse can be avoided, long-term over-exploitation of the stock is good neither for the species and ecosystems nor for the fishers who depend on them.

A second key source of difference is the pattern of ownership of the resource in fisheries. With most resources, one legal entity (a person or a company) owns the rights to a particular resource. A farmer either owns or has exclusive right to farm a piece of land, a mining company either owns or has exclusive right to mine a particular site. Fisheries work differently. With the exception of the exploitation of a few sedentary species like mollusks, no one has exclusive rights over a particular fish. Individuals may own quotas, which are the right to fish a certain amount. But even then, the quota is the right to a certain proportion of the stock, not the right to particular individual fish (and in most cases not even the right to a particular amount of fish over the long term). This structure generates a clear collective action problem. The problem is clearest in international fisheries, where there is no sovereign authority. But even at the national level, when a specific government has clear authority to regulate fisheries, the fish remain unowned. Because ownership, in the absence of regulation, is key to the functioning of markets, the market for fish cannot self-regulate, cannot balance supply and demand in a way that preserves stocks for the future. Authoritative regulation is necessary.

A third key source of difference is that global patterns of industry organization are diffuse. Many resource industries are dominated by a few key companies, or by governments that control the resource directly. Petroleum is a good example – a handful of huge multinational corporations dominate the industry, as do the governments of the countries with the biggest reserves. In agriculture, a few companies control much of the world's seed supply. There are no equivalents in the geopolitics of fish. Instead, for the most part, there are national industries (often made up of many smaller-scale fishing companies or individual fishers) able to extract subsidies from their governments. Even the international distribution and processing of fish, while more concentrated than fishing itself, remains relatively diffuse. Governments thus cannot intervene effectively in the management of global fisheries by regulating or pressuring a few large companies. Regulation must directly affect the behavior of a wide range of small companies and individual operators.

To this point, for the most part, governments have not regulated effectively. Some national fisheries, such as Iceland's, are managed more or less sustainably, and other national fisheries, such as the United States', are managed much better than they used to be. Some disasters with individual fish stocks, such as Patagonian toothfish, have been prevented. But on the whole, the world's fisheries are under-regulated and over-fished. Limited successes are often more than offset by abject failures, such as efforts to protect bluefin tuna. Aquaculture can take some of the pressure off of the world's capture fish-eries, but it can at the same time increase that pressure by maintaining the fiction that we have a limitless supply of fish.

Some newer forms of regulation, such as the creation of marine protected areas or the allocation of catches (most often as percentages of quotas, so the amount increases or declines as the overall allowed catches change) to individuals as property rights that can be traded have shown some signs of success. Additionally promising approaches, discussed in this chapter, include efforts to change the behavior of fishers or of regulators through changing the preferences of consum-ers. Ultimately, though, there need to be fewer people in the fishing industry for fish to become a sustainable resource.

This chapter begins with an overview of efforts to change consumer preferences and, via collective action of consum-ers, put pressure on those harvesting or farming seafood to do so in a more sustainable manner (or reward those who do). Lists, certification, and collective action such as boycotts or buying commitments, can change preferences and incentives. The chapter then wraps up the major discussions of the book, re-evaluating the future of the world's fisheries and making specific suggestions for consumers.

Consumer-Driven Measures

Governmental and intergovernmental regulatory processes have serious shortcomings in addressing overfishing. The political power of fishing interests within countries makes national governments reluctant to restrict fishing in national waters to the extent that fishery health requires. The voluntary nature of international regulation means that unless all states that license vessels to fish in an area or for a species are willing to accept and monitor restrictions for their vessels (when vessel owners do not, for the most part, want to be regulated) it can be extremely difficult to reach international agreement to limit catches or technology.

But there are non-regulatory approaches that can have an effect on fishing behavior, primarily through influencing the preferences and behavior of consumers of fish. One of the most successful examples of consumer pressure was in the late 1970s through early 1990s, when consumer pressure, primarily in the United States, the largest market for canned tuna, changed the way in which some species of tuna were caught, as noted in Chapter 2. At the beginning of that period, yellowfin tuna, including those that ended up in cans in supermarkets, were caught by encircling dolphins (with which they school) in purse-seine nets. As many as 300,000 dolphins were killed in this manner per year in the early 1970s, with more than 100,000 killed by the US fleet alone. Consumer pressure led the US government to pass laws requiring that American tuna fishers operate in a dolphin-safe (or at least dolphin-safer) way, and attempted to force other countries to adopt similar rules. This attempt fell foul of international trade rules, but consumer pressure ultimately worked at least as effectively through the supply chain to change fishing behavior.

As the public became aware of dolphin deaths from tuna fishing in the 1980s, at least in part due to a documentary produced by Earth Island Institute, environmental organizations in the United States led a consumer boycott of canned tuna and called for an end to fishing practices that harmed dolphins. Heinz, the company that sells tuna under the Starkist brand, decided to brand its tuna as dolphin-safe, a practice quickly followed by the other two major canned tuna companies. In early 1990 these three major tuna brands announced that they would no longer buy or sell tuna for the US market if it was caught in ways that killed dolphins. With the American market for canned tuna, by far the world's largest, mostly closed to tuna caught in the traditional way, the global price difference between dolphin-safe and dolphin-unsafe tuna became greater than the additional cost of catching tuna in a dolphin-safe way.

Ultimately governments representing the major tuna-catching countries did cooperate to create international rules that work to decrease the traditional dolphin-unfriendly way of catching tuna. This cooperation has been quite successful, with global dolphin mortality from tuna fishing falling to a current level of below 5,000 annually. (It is not a panacea, however; other forms of bycatch have increased.) This cooperation was only made possible because the market had already forced a change in behavior by the world's industrial-scale tuna fishers, through the mechanism of consumer pressure. Forcing this kind of change is not easy, despite the evidence of success in the tuna case. This example involved changing fishing practice to decrease one type of incidental damage caused by fishing, rather than reducing the amount fished. As such, it did not have as big a direct effect on the livelihood of fishers as a smaller catch would. But it is an example of consumer pressure resulting in changed behavior by fishers, and in a positive environmental impact.

As in the tuna–dolphin example, other consumer-driven efforts to change fishing behavior usually need some sort of organizing mechanism to give them traction. There are several different approaches that have been taken to organize the power of consumers to put pressure on fishers or elsewhere in the supply chain, to increase the sustainability of fishing. These include the creation of lists, the certification of sustainable fisheries, and collective action strategies (such as boycotts or buying agreements) to increase the reach of the lists and certification systems. These strategies share a similar focus on the provision of information that then allows consumers, or other actors, to make decisions about which seafood to buy or sell. These strategies interrelate – a boycott may be called of fish that are listed as threatened, or retailers or restaurants may decide only to purchase fish that have been certified as sustainable. These approaches may also drive pressure for national or international regulation. It is worth examining what each strategy contributes.

Lists

Some organizations provide lists of which types of seafood to consume or avoid based on either its sustainability or its health risks. The best known of these is the one put out by the Monterey Bay Aquarium, which creates a Seafood Watch pocket guide (the guide is now available in versions for different regions, with an additional one focusing on sushi). It lists those fish that are "best choices," or "good alternatives," as well as those to "avoid." The Aquarium began its list in 1999, and updates and expands it regularly. It has distributed tens of millions of pocket guides and its iPhone app has received more than 70,000 downloads. It created an organization, Seafood Watch, to oversee its listing process. This process is transparent (and peer-reviewed), and involves gathering and evaluating information on the status and trends in the stock, paying particular attention to the vulnerability of the stock to fishing pressure, the current status of the stock, the extent of bycatch involved in fishing operations, the harm that fishing practices do to ecosystems and the effectiveness of the current management regime for that fishery. The organization has a similar set of criteria for aquaculture operations.[1]

Other organizations create similar lists. The Environmental

Defense Fund also has a three-part list ("eco-best," "eco-OK," and "eco-worst"). The Audubon Society for years provided a wallet card giving information about the background, status, management, and bycatch and habitat implications of consuming particular fish, rating their impacts. Audubon has recently ceased this practice and now directs consumers to Seafood Watch. Greenpeace International creates a "red list" of fish species to avoid buying in the supermarket. The Australian Marine Conservation Society and the Royal Forest and Bird Protection Society of New Zealand both publish guides to the best fish to eat from a sustainability perspective. Others have proliferated in the past decade.

It should be noted that each rating creates an index, meaning that a rating system represents an amalgam of factors. The aspects organizations include in their rating systems are such things as the extent to which individual species or populations are being overfished, the side-effects of fishing or farming a species (such as pollution) and the health effects of the fish. If a person is more concerned with one of these factors than the other, more individual research is necessary. The consumer may rank the importance of these factors differently than the listing organization does.

These kinds of efforts are intended to influence consumer behavior. The fact that the Monterey Bay Aquarium distributes its guide in a size meant to fit into a pocket suggests that consumers are supposed to carry it with them (or use it on smartphones). Other guides make themselves similarly useful.

There are also organizations that create lists of seafood to seek out or avoid based on health concerns. PCBs and heavy metals (such as mercury) are common contaminants in fish that are high on the food chain, as discussed in Chapter 1, and many of the most popular fish species are sufficiently contaminated that some organizations recommend that consumption be limited or avoided altogether. The Environmental Defense Fund Health Alert chart suggests that some wild species, such bluefish, and sturgeon, be avoided altogether, and others (such as bluefin tuna or king mackerel) be eaten only rarely.[2]

The idea behind these lists is that consumers will, on their own, decide to stop consuming the seafood that is overharvested; there will thereby be less demand for these species and fishers will earn less profit by catching them and will therefore focus on catching fish in a less environmentally harmful way. It is difficult to determine the extent to which these lists have changed consumer behavior or, more importantly, whether changed consumer preferences have affected fishing practices. The vast numbers of lists distributed and downloaded certainly suggests that consumers pay some attention to these issues, but no one has been able to measure the collective impact of these individual decisions.

Certification
Rather than relying on governments to pass regulations, non-governmental entities (or, depending on the issue, even governments) may decide to certify something as environmentally preferable. This process is similar to the lists described above, in that an organization designates something as environmentally good and publicizes that designation. The difference in part is the process: certification creates a transparent standard against which any seafood operation can be evaluated to determine its sustainability; the certification process creates a benchmark that is either met or is not, determined by an evaluation by an independent organization.

This approach is often called private regulation (to contrast it with public – governmental – regulation). It is also seen as a form of (eco-)labeling, since a product certified by a private entity or organization bears a label that declares it thus certified. The process of creating standards and evaluating operations to determine whether they meet those standards is done in a non-regulatory way; no one is required (at least not by the private regulatory body) to meet the standard, but producers can choose to do so.

The primary certification for seafood is done through the Marine Stewardship Council (MSC), an organization begun in 1997 in a collaboration between the non-governmental organization WWF and the major food manufacturing conglomerate Unilever, although it has since become an independent organization. The organization sets standards for what constitutes a sustainable fishing operation. Such operations must keep fishing at a sustainable level for a given stock, must minimize environmental impact, and manage the fishing effectively, including meeting all existing laws at the local, national, and international levels. Certification is for a set of actors, fishing a particular species in a designated area, using a designated technology.

As of mid-2010 there are eighty-nine MSC certified fisheries. Another 120 are in the process of undergoing certification, suggesting that the popularity of the program is increasing. Certification is good for five years, and certified stocks earn the right to use the organization's label in publicity and marketing materials. There is also a chain-of-custody certification (lasting for three years) that distributors can receive to certify that seafood they sell (including processed seafood, which may involve fish from many sources) comes entirely from certified stocks.

One main advantage of certification is that it addresses the supply chain more broadly, whereas positive and negative lists primarily work through consumer demand. When it works well, a certification system changes behavior in multiple places along the supply chain. For instance, the New Zealand hoki fishery changed its management plans and put into place policies to avoid bycatch of seals in order to achieve MSC certification. If those who make use of the fish as inputs (because hoki serves the bulk global fish market) achieve chain-of-custody certification, they will be required to source all their inputs from certified fisheries, thus expanding the reach of the program.

There are some criticisms, however, of how the MSC operates. Some specific certification decisions – British Columbia sockeye salmon, Alaskan pollock, and also the New Zealand hoki – have been criticized by conservation organizations because of declining stock estimates in those fisheries. The system the MSC uses to invite objections to certification applications rarely, if ever, results in a rejection. The complexity and cost of getting certified also means that most certified producers are from developed countries. And the MSC model also is more amenable to certifying industrial, rather than artisanal, fishing, even though the latter may be more likely to be sustainable.

There are other competing certification standards. The Swedish organic certification agency, KRAV, certifies fisheries and is a well-known standard in Scandinavia. The Japanese government is creating a fishery certification process, although it is likely to be less stringent than many existing ones. The increasing popularity of the MSC process, and its geographic reach both in terms of catches and markets, suggests that it is likely to become the predominant certification body.

Consumer Collective Action – Boycotts and Buying Commitments

Although both lists and certification can have some effect on consumer behavior, a more direct way for consumer preferences to translate into market effects that change fishing behavior is for other parts of the seafood supply chain to exhibit preferences for sustainably harvested seafood. This approach involves groups acting collectively to make use of, and build on, the information the listing or certification processes provide. An early effort of this type was a campaign that began in 1998 to "Give Swordfish a Break." The campaign, organized by the non-profit environmental organizations Seaweb and the Natural Resources Defense Council, persuaded major US chefs (more than 700 of them) to boycott swordfish because it was being inadequately regulated by the International Convention for the Conservation of Atlantic Tunas (ICCAT) (which regulates swordfish catches in addition to tuna). Largely as a result of the publicity generated by the boycott (although also because of new availability of other species, such as Chilean sea bass), prices fell and states agreed, perhaps because of the lower profits, to greater protections under ICCAT. The boycott was called off in 2000.[3]

A similar restaurant chef effort on behalf of Chilean sea bass (Patagonian toothfish) was organized by the National Environmental Trust (a US-based environmental organization) in 2002. The fish, discussed in Chapter 1, rose to sudden popularity, especially on restaurant menus in the United States and Europe, in the 1980s. It is managed internationally by the Commission for the Conservation of Antarctic Marine Living Resources (CCAMLR), but the organization faced a major problem of fish caught outside its regulatory process. A key problem in managing the fishery was that many of the ships that fished the species were flagged in countries that were not members of CCAMLR. After years of failing to persuade non-member states to join the agreement, CCAMLR members eventually agreed to prohibit the importation of Chilean sea bass that has not been caught within CCAMLR's regulatory process. The organization set up a Catch Documentation Scheme by which fishers had to certify the origin and rule-adherence of their catches of Chilean sea bass, and member states agreed to only land, or transship, fish with the relevant documentation. Although fish can still be caught by non-member-flagged vessels, they fetch a much lower price, because the primary CCAMLR member states are also the primary consumers of the fish. The fishery may not yet be sustainable, but it is at least better managed. In this case, the link between consumer pressure and outcome is somewhat diffuse; governments may have become concerned about overfishing in Antarctic waters anyway. But it may well be the case that popular concern encouraged this process, and led to a stronger agreement, and one reached more quickly, than would otherwise have been the case. In this way consumer pressure may play a role through diffuse channels.

One of the major downsides of boycotts is that they are a blunt instrument; they do not differentiate based on where a fish is caught, or how. They may be useful in persuading people to avoid a fish that is so overfished as to be endangered, but they have less of an effect in changing the behavior of fishers toward more sustainable fishing practices, or in rewarding fishers that fish a given species in a more sustainable manner than others.

The scale of the point of contact matters in terms of the effects of consumer collective action – big retailers and restaurant chains can be individually successful in demanding changes in the behavior of fishers, while small restaurants and retailers do not have the market power to do so. In aggregate, however, they may be able to have some impact. It is also the case that the types of fish on which these boycotts can have an effect are the category of global differentiated fish discussed in Chapter 3. Consumer pressure is more likely to impact restaurants than food processing companies, since Chilean sea bass or bluefin tuna is much more likely to appear on the menu in expensive restaurants than in cans on supermarket shelves.

But similar collective pressure can be exercised through retail outlets or mass-produced seafood that can have an effect on the fish supply chain in similar ways. This pressure may be exercised less through a boycott of unsustainable fish than through a decision to source sustainable, or certified fish.

For example, Walmart, the world's largest retailer, announced in 2006 its intention to have all the wild fish it sold in North America be MSC-certified by 2011. (Because MSC certification does not, at this point, apply to aquaculture, Walmart will still sell non-certified aquaculture products.) Other major retailers have made similar promises; in Canada, the grocery chain Loblaws vowed to source all its wild seafood from MSC-certified fisheries by 2013.

Another avenue for sustainability purchasing decisions to reach a broader audience is fast-food restaurants. Here the effect is possible on the bulk global supply of fish, rather than just the high-profile species sold in supermarkets or restaurants. McDonald's fish sandwich is made out of whitefish, but processed in such a way that it does not matter which whitefish specifically are used. Beginning in 2001, McDonald's worked with the environmental organization Conservation International to increase the sustainability of its fish purchases, based on a rating system of conservation health. The fast-food chain has shifted its purchases of whitefish based on sustainability standards and boasts that only nine percent of its fish purchases in 2007 came from fisheries deemed unsustainable. In 2009 its director of supply chain management was awarded the Seafood Choices Alliance Champions Award in recognition of the focus on sustainably sourced fish. McDonald's is a large enough purchaser of whitefish that its purchasing decisions can have an effect on the global market and on fishing behavior. For example, suppliers of cod from the Baltic sea tightened their reporting standards after

McDonald's stopped buying fish from that region because of insufficient reporting.[4]

Decisions by restaurants or stores to source only certified seafood takes decisions about sustainability of seafood purchases out of the hands of the consumer and thereby has a much further reach. If the store at which you buy your fish only sells certified seafood, you are not even given the choice to purchase fish from threatened stocks. Those types of policies extend the reach of certification processes, because they do not depend on an educated, aware, or concerned consumer. Even though consumer awareness is valuable, a strategy that relies on all individual consumers making environmentally aware seafood purchasing decisions is unlikely to succeed. Although consumer pressure initially may lead to these larger collective decisions to sell only sustainable fish products, taking the decision away from the consumer increases the number of people who ultimately eat only certified fish when they eat fish.

Consumers have a close connection to the actual fish catch, whether they realize it or not. They are the ones who will notice when fish is more or less available and who will experience the dangers if the fish they would like to eat is contaminated with toxins; their purchasing behavior, even if they are not prioritizing sustainability, will influence fishing behavior. And although consumer action can be – and has been – important in changing fishing behavior by individual fishers or large conglomerates, consumer behavior must be seen within the broader context of the fishing industry and governments in ultimately determining the future of the world's fisheries.

The Future of Fisheries

Are the world's fisheries on a path to disaster? Current global patterns of fishing, and of aquaculture, are unsustainable.

A sudden and complete collapse is unlikely, because some national fisheries are managed better than others, and some species are exploited more thoroughly than others. But as some fisheries are depleted or collapse, the pressure will only increase on others. Furthermore, many contemporary fishing technologies do huge damage to the marine ecosystem, even beyond the target species of the fishery. Some of these technologies, such as purse-seine nets and longlines, are less prevalent than they used to be. But others, such as trawling, continue to be as popular as ever. Trawling, when done at the bottom of the ocean, can scrape acres of sea floor bare on each pass by a large ship, and much of the marine life caught is rejected and wasted and most likely dies. The damage, in other words, extends well beyond depletion of the species that fishers are targeting.

Aquaculture presents a different set of environmental challenges. It is not limited in scale as rigidly as capture fisheries are, and much of the damage done by inadequate regulation is in the form of local pollution. But aquaculture can create some real dangers both to capture fisheries specifically and to marine ecosystems and the marine environment more generally. Aquaculture of carnivorous fish increases overall pressure on the global capture fishery, and poorly regulated aquaculture more broadly can spread both invasive species and disease to marine ecosystems.

But the problems confronting the world's fisheries can be addressed, and the world's fisheries can be managed much more sustainably than is currently the case. Doing so effectively requires that we approach the problem from three perspectives. First, we need to improve on current patterns of fisheries management by the world's governments. We know more about how to make government regulation of fisheries work than we used to, and there is considerable scope for improvement in the current system. Second, we need to change the way we think about subsidizing fishers. If we continue to subsidize the fishing industry, fishers will continue to fish too much. And third, we need to be informed and conscientious in the decisions we make about our own consumption of seafood. Your dinner choice tonight may not have much overall impact on the state of the world's fisheries. But the collective choices of consumers can have a major impact on what gets caught and how it gets caught, both through the market directly and by making political decisionmakers aware of popular concern about the sustainability of the world's fisheries.

Managing Fisheries

A first step in managing the world's fisheries more sustainably is improving the systems we already have in place. States manage the waters in which most commercial fishing is done, and they have the authority to compel compliance with rules. Domestic rules will work best if states do not treat their EEZs either as sources of new economic development or as open access (even if only to their national fishers) resources. States also need to be careful of the unintended consequences of prioritizing support to their artisanal fishers in ways that can increase the intensity of that fishing sector.

At the international level, the existing regulatory structure is based on regional fisheries management organizations. These organizations set catch limits (and other regulations) their member states need to follow. But because the limits are passed by states concerned about the short-run impact on their fishers who prefer not to be unduly restricted, the limits are frequently set much higher than the organizations' own scientific commissions recommend. States offering ship registration to those wishing to escape international rules give ships a legal loophole for avoiding international regulation, and so additional fishing is done by fishers not bound by the already-too-high international limits. The decisionmakers within RFMOs need to be willing to restrict catches to levels advocated by scientists. They need to create incentives (as CCAMLR has done, with member states refusing to land or transship toothfish caught outside the regulatory process) for fishing vessels to register in RFMO member states and follow the rules. If political leaders received pressure from those interested in conservation (and not just those interested in unlimited fishing) they might be more willing to agree to international limits that, in the long run, will help fishing continue to be an option indefinitely.

These organizations also need to work together cooperatively to address the underlying problem of regional- and species-based management: the shift of fishers to other species or regions in response to regulation. Although some RFMOs have had some success protecting some species, the overall health of global fisheries has not improved, and as long as the number of fishers fishing globally remains the same (or increases), along with increasingly efficient technologies for finding and catching fish, the piecemeal approach to global fisheries regulation cannot ultimately protect global fish stocks. Some RFMOs, like those protecting various species of tuna in different regions, have begun efforts to collaborate, which is a hopeful sign. But more needs to be done to make international regulation seamless and compatible across regions and species, and to decrease the amount of fishing overall.

Along with RFMOs, other international institutions could potentially be useful in managing global fisheries, and in particular in helping with the most threatened species. In some cases unilateral action by states that constitute the major markets for fish, particularly high-value, globally traded species, may also prove effective. The effort by the United States, beginning in the 1970s, to save dolphins from being killed in tuna-fishing nets, is one example of effective unilateralism.

Another effort by the United States to save sea-turtles being killed in shrimping nets was not entirely successful, but was at least partially so. It also resulted in a ruling by the World Trade Organization's Dispute Settlement Mechanism, often referred to as the world's trade court, that states could use unilateral market power to try to save species to support a multilateral goal, if efforts to deal with the problem multilaterally were tried and failed.

An example of other international institutions being used to impact fisheries practice is a recent attempt by several countries to have bluefin tuna listed as endangered under the Convention on International Trade in Endangered Species of Wild Fauna and Flora (CITES), which would prohibit international trade in the species, and thereby limit pressure on the most endangered stocks in the Atlantic Ocean and Mediterranean Sea. This proposal was a conscious attempt to do an end run around the International Convention for the Conservation of Atlantic Tunas, which is supposed to regulate the catch but seems unable to prevent serious overfishing. This attempt failed, because many countries voted against the measures specifically to prevent a precedent of moving fisheries regulation to the most restrictive international organization. But continued attempts may well eventually succeed. And listing fish species as endangered under CITES can reduce the pressure on them whether they are fished in international waters or are entirely under the regulatory authority of one country, because the most endangered fish are often also those that are most traded internationally.

Beyond using existing patterns of international cooperation more effectively, two techniques that have proved highly effective in managing fisheries at the domestic level in some countries might be both promoted globally to countries not already using them, and tried at the international level. The first of these is individual transferable quotas, or ITQs. Where implemented comprehensively, particularly in Iceland and New Zealand, these have proved highly effective at limiting fishing effort to sustainable levels. The creation of an international system of ITQs would likely have a major impact on the effectiveness of international fisheries management. There are considerable political obstacles to creating such a system, and it is not likely to appear in the near future. Nonetheless, it should not be written off – efforts to build some political momentum toward an international ITQ system could help to overcome those obstacles.

The second technique is the creation of marine protected areas (MPAs), or the even more restrictive marine reserves. These are areas where fishing and other activity is severely restricted, and in some cases no fishing at all is permitted, allowing both individual species and marine ecosystems more generally a haven from constant human depredation. Research has shown that MPAs not only are effective as insurance against the collapse of fish stocks, but they also make surrounding fisheries considerably more productive. Creating more MPAs within national waters and EEZs, and finding a way to create MPAs in international waters, would decrease the risk of ecosystem collapse, and possibly increase the productivity of the world's fisheries overall.

Seeing Fishing as an Industry

Improving on current regulatory patterns, however, will by itself be insufficient to ensure sustainable management of the world's fish resources. The level of pressure on these resources from the fishing industry is greater than can be contained by rules designed to protect particular species or particular areas of water, because there is simply too much fishing capacity out there. Not only is the industry bigger than global fish stocks can sustain, it is considerably bigger than the global market for fish can sustain, as indicated by the huge losses that the industry runs, year after year. It exists at its current size only because of government subsidies, and these subsidies generally work at direct cross-purposes to fisheries regulation. The romance and history of fishing is not a sufficient argument for continuing to spend money on fishers the global stock of fish cannot support. Eliminating, or even lowering, these subsidies will allow regulation to work more effectively.

There are three general reasons why the fishing industry is so heavily subsidized. The first is the same reason that industries in general can often attract government subsidies. Industries are concentrated, and it makes economic sense for them to invest in extensive lobbying of the government for special privileges. Taxpayers and consumers (and environmentalists in this case) are a more diffuse group, and are therefore less likely to lobby effectively to prevent rules that cost taxpayers and consumers money, and harm the environment. This phenomenon is seen in the regulation of many industries, though, and does not explain why fishing is more heavily subsidized than most industries.

The second reason is that governments often see fishing as a convenient route to economic development. Subsidizing the development of the industry takes advantage of a free-access resource, and there will always be a market for fish (even if it only pays below cost). Fishing can also be practiced at a wide range of levels of technology and capital intensiveness, making it a flexible venue for government support. Using fish resources as a route to economic development often fails, because of the fixed limit on available fishing resources. Unless the development of the industry is restrained at well below the level at which it reaches the carrying capacity of targeted fishery stocks, it is likely to overshoot that level. If not, improvements in productivity over time will push fishing capacity above carrying capacity. The government, having invested resources in building the industry, will then have to invest further resources in restraining it.

The third general reason that the fishing industry is so heavily subsidized, and the one that creates perhaps the biggest obstacle to reducing subsidies, is its social image. Fishing is often seen not just as an economic venture, but also as a romantic undertaking and a part of local culture. As such, it is often subsidized for cultural and historical as well as (or rather than) purely economic reasons. But fishers do not in fact want to maintain traditional fishing cultures, which generally left most fishers in dire poverty. Subsidies are an attempt to maintain employment in fishing in areas where it is a traditional pursuit, while avoiding the poverty. But this approach ultimately backfires – standards of living go up, fishing productivity goes up, but the fish stock stays the same size or shrinks. Governments therefore need either to continually increase their subsidy of the industry while desperately trying to keep it from fishing as much as it can, or to allow (or, ideally, encourage) the industry to shrink.

Not all subsidies are necessarily bad. Governments can play a useful role in actively assisting the shrinking of the industry, by buying fishing vessels to take them out of circulation, or by paying people already in the industry to get out. Beyond that, however, subsidizing the industry for cultural reasons merely puts off the inevitable. We have a choice between shrinking the industry now, or doing so later when global stocks are even smaller. And doing so now requires that we recognize that maintaining traditional fishing cultures, particularly in developed countries, is simply not a long-term option. As currently practiced in developed countries, fishing is in any case a heavy industry, far removed from the romantic image of the sea often portrayed in popular culture. Subsidizing the fishing industry undermines environmental sustainability at sea without creating socioeconomic sustainability in fishing communities. Promoting environmental sustainability at sea requires recognizing that there is a clear link between over-fishing and over-employment in the industry. And it requires at minimum reducing, and ideally eliminating, subsidies.

All of these mechanisms for improving the state of the world's fisheries, both by improving regulation and decreasing subsidies, require that governments take an active interest in sustainable fisheries management, both individually and cooperatively. It often requires that they do so against the active opposition of the fishing industry, which is sometimes more concerned with maintaining short-term income, and with competing with foreign fishers, than with the long-term effects of overfishing. In order reasonably to expect governments to take on this role, they need active political support from constituencies outside of the industry. A key role for concerned citizens therefore is to make their concerns known to their governments, either directly or through support of non-governmental organizations that are involved in the issue. These will usually be environmentally focused NGOs (like Greenpeace, the Environmental Defense Fund, Conservation International or the International Union for the Conservation of Nature (IUCN)). Generating public pressure on governments to act is a key element in helping to prevent continued overfishing globally. But individuals can have an impact on fisheries practices directly, as well, though the decisions that they make in purchasing and eating seafood.

Eating Sustainably
Where we as individuals can have the most impact on the health of the world's fisheries is in our choices of seafood to eat. Some ideas about what to eat are straightforward. Reduce waste – don't buy more than you need, don't cook more than you need. Don't feel a need to eat seafood every day. Be wary of generic 'fish' products, like fish sticks, fingers, or sandwiches,

that lack any indication of what is inside, or where it came from. These products are usually made from the cheapest possible sources of marine protein, which are unlikely to also be the most sustainable sources. When purveyors go to the effort of finding sustainably caught generic fish, they will usually let you know.

Other ideas can be gleaned from earlier chapters in this book. Fish from deeper seas tend to be slower-growing, and so should be eaten sparingly (Patagonian toothfish/Chilean sea bass serves as a good example here). The same is true of trawled fish, and many non-farmed shrimp, because of the damage that trawls do to the sea floor and the bycatch involved. Farming carnivorous fish increases pressure on capture fisheries, and so farmed carnivorous fish (such as salmon and shrimp) should also be eaten sparingly if at all. And, of course, endangered species (such as bluefin tuna or many kinds of shark) should be avoided altogether.

But the choices are not always obvious. Farmed salmon is often marketed as an environmentally preferable alternative to declining stocks of wild Pacific salmon. But in ways the wild salmon are the environmentally preferable choice, if the concern is for global fish stocks. On the other hand, caviar from farmed sturgeon is vastly preferable to caviar from sturgeon caught in the wild, because all wild sturgeon are endangered. Sometimes local is better, sometimes it is worse (and sometimes it is implausible – if a restaurant in Florida has local grouper on the menu, odds are good that it is a lie, and that it is some other kind of fish altogether, because the commercial grouper fishery in the area is closed).[5] Most experts, who can devote all of their working time to keeping track of these issues, and of the rapid changes both in the condition of various stocks and the ways in which they are caught or farmed, have trouble keeping track of everything. How can individual consumers do so?

There are two ways of doing so, by eating fish of a general kind that is less problematic, and by eating fish that have been individually certified as sustainably harvested or grown. The first route relies on organizations, discussed at the beginning of this chapter, that keep track of global fisheries, and report on overall patterns with respect to specific kinds of fish, and even specific fish from specific regions. These organizations may also at the same time track health issues with respect to particular types of fish, noting for example which fish have high levels of useful omega-3 oils, or high levels of harmful mercury accumulations. Checking a Seafood Watch pocket guide or the Environmental Defense Fund's Health Alert every now and then can help you keep up to date on what is becoming more sustainable, and what is becoming less so.

And, as noted above, you do not always get the fish you think you are getting. Anything sold simply as "fish" (including things like "imitation crab") could really be anything. Large companies with national reputations to be concerned about, such as major (and particularly upscale) supermarket chains, will likely be careful about selling you what they claim to be selling you. Tourist-oriented restaurants that probably do not expect you back anyway, on the other hand, are less likely to expect anyone to look into what they sell, and are less likely to care anyway.

The second way of making sure that the fish you eat are relatively sustainable is to buy fish that are certified as being caught or raised sustainably, as also discussed at the beginning of this chapter. Certification means that the fish from the region it comes from, and in the manner it was caught, has met a specific set of criteria, and that the specific fish you are buying can be guaranteed to have met those standards. The Marine Stewardship Council's certification program is likely to continue to be the predominant one in this area.

The best-known certification system for food in general is

organic standards (and the equivalent or even stricter standards in most other developed economies, often referred to in Europe as "biodynamic" rather than "organic"). In the United States, the standards are set by the federal government (through the US Department of Agriculture), but are monitored by a set of Accredited Certifying Agents (which may be businesses, NGOs, or parts of state governments). "Organic" standards in the United States do not apply to capture fisheries and they do yet apply to aquaculture, although standards are in the process of being formed, and some operations or sellers label their aquaculture fish as organic. The nascent standards cover what fish are fed, and what chemicals and medicines are used in raising them, so they are not relevant to fish caught in the wild. The standards will, once formally adopted, provide useful information about farmed fish. Organic fish, for example, should have been fed mostly organic feed, and have not been heavily dosed with antibiotics or other medicines. As such, these fish are probably better for the environment (and for the eater) than non-organic, other things being equal. But these particular standards are not designed to address the broader impact of aquaculture on the environment, and are not designed to deal with caught fish (still the majority of seafood produced) at all.

Final Thoughts

This book began with a consumer decision. It tracked a Patagonian toothfish in the Antarctic Sea, on its way to becoming Chilean sea bass on a restaurant plate. The story of this particular fishery is a good example both of our abuse of the world's fishery resources, and of the potential for a more sustainable relationship between humanity and its seafood. A decade ago Patagonian toothfish were being desperately overfished. The international body charged with managing the fishery, the Convention for the Conservation of Antarctic Marine Living Resources (CCAMLR), was aware of the problem and trying to do something about it, but was having little success. Now, however, management efforts are mostly working. The species is probably still overfished, but not catastrophically, and mechanisms are in place to reduce catches if the stock starts showing signs of collapse.

What changed? Two key things. First, the fish became famous. It became a *cause célèbre*. Environmental organizations put it on their agendas, and big-name chefs took it off their menus. Decisionmakers noticed, and this gave impetus to efforts to manage the fishery multilaterally. Second, governments of the major consumer countries, such as the United States and the countries of the European Union, became more willing to deal aggressively with the problem of fishing outside of international quotas. They imposed a licensing system to ensure that any Patagonian toothfish that came into the country was caught within, rather than outside, the quota system. This system undermined the market for non-quota fish. It did not eliminate the problem, but it significantly diminished it.

This example suggests two routes by which we can improve our collective stewardship of the world's fisheries, individual eating decisions and political activity. But it also suggests that changing our current pattern of abuse of the world's fisheries will not be easy. Patagonian toothfish is a high-value fish caught in a remote place by large ships, making it relatively easy to monitor. It has a recognizable brand-name and is eaten mostly by relatively wealthy people in restaurants, making it relatively easy to target a publicity campaign. Even so, current fishing is, at best, at the limits of the species' carrying capacity.

Notes

The authors would like to thank Shilpa Idnani and Cassandra Palanza Gaudin for research assistance, Marsin Alshamary and Sarah Oddie for assistance with the index, Louise Knight and David Winters at Polity for both suggesting this project and shepherding it through the publication process, and two anonymous reviewers for helpful comments on an earlier draft of the manuscript.

1 INTRODUCTION

1 FAO, *The State of World Fisheries and Aquaculture* (Rome: FAO, 2009), pp. 58–65.
2 Boris Worm et al., "Impacts of Biodiversity Loss on Ocean Ecosystem Services," *Science* 314(3 November 2006): 787–90.
3 FAO, *The State of World Fisheries and Aquaculture* (Rome: FAO, 2009), p. 9.
4 Ransom A. Myers and Boris Worm, "Rapid Worldwide Depletion of Predatory Fish Communities," *Nature* 423 (2003): 280–3.
5 Garrett Hardin, "The Tragedy of the Commons," *Science* 162(3859) (1968): 1243–8.
6 Daniel G. Boyce, Marlon R. Lewis & Boris Worm, "Global Phytoplankton Decline Over the Past Century," *Nature* 466 (29 July 2010): 591–6.
7 Matteo Milazzo, *Subsidies in World Fisheries: A Reexamination*, World Bank Technical Paper No. 406 (Washington, DC: The World Bank, 1998); Ussif Rashid Sumaila and Daniel Pauly, eds, *Catching More Bait: A Bottom-Up Re-Estimation of Global Fisheries Subsidies*, University of British Columbia, Fisheries Center Research Reports 14(6) (2006).

2 GROWTH OF THE GLOBAL FISHING INDUSTRY

1 FAO, *The State of World Fisheries and Aquaculture 2008* (Rome: FAO, 2009).
2 Mark Kurlasky, *Cod: A Biography of the Fish that Changed The World* (New York: Walker and Company, 1997).
3 Callum Roberts, *The Unnatural History of the Sea* (Washington: Covelo Press, 2007), p. 185.
4 Reg Watson and Daniel Pauly, "Letter: Systematic Distortion in World Fisheries Catch Trends," *Nature* 414 (29 November 2001): 534–6.
5 Callum Roberts, *The Unnatural History of the Sea* (Washington: Covelo Press, 2007).
6 FAO, "Fishing Vessels," http://www.fao.org/fishery/topic/1616/en.
7 Elizabeth R. DeSombre, "Fishing Under Flags of Convenience: Using Market Power to Increase Participation in International Regulation," *Global Environmental Politics* 5(4) (November 2005): 73–94.
8 FAO, *The State of World Fisheries and Aquaculture 2008* (Rome: FAO, 2009), p. 13.
9 FAO, *The State of World Fisheries and Aquaculture 2008* (Rome: FAO, 2009), p. 7.
10 Elizabeth R. DeSombre, *Domestic Sources of International Environmental Policy: Industry, Environmentalists, and U.S. Power* (Cambridge MA: MIT Press, 2000).
11 Rupert Sievert, " A Closer Look at Blast Fishing in the Philippines," *OverSeas* 2(5)(May 1999), http://www.oneocean. org/overseas/may99/a_closer_look_at_blast_fishing_in_the_ philippines.html.
12 Richard Ellis, *Tuna: A Love Story* (New York: Alfred A. Knopf, 2008), p. 140.

3 STRUCTURE OF THE FISHING INDUSTRY

1 FAO, *The State of World Fisheries and Aquaculture* (Rome: FAO, 2009), p. 26.
2 From UN FAO, "Artisanal Fishing: Definition," http://www.fao. org/fi/glossary/default.asp.

3 Gerry Kristianson and Deane Strongitharm, "The Evolution of Recreational Salmon Fisheries in British Columbia," *Report to the Pacific Fisheries Resource Conservation Council* (Vancouver, BC: Pacific Fisheries Resource Conservation Council, June 2006).

4 Figures are from OECD, *Review of Fisheries in OECD Countries 2009: Policies and Summary* (Paris: OECD, 2010).

5 Poseidon Aquatic Management Ltd, for Pew Environment Group, "FIFG 2000–2006 Shadow Evaluation" (March 2010).

6 UNEP, *Analyzing the Resource Implication of Fisheries Subsidies: A Matrix Approach* (UNEP 2004).

7 K. Dahou, and M. Deme, with A. Dioum, "Support Policies to Senegalese Fisheries," in UNEP, *Fisheries Subsidies and Marine Resource Management: Lessons Learned from Studies in Argentina and Senegal* (Geneva: United Nations Environment Programme, 2002).

8 UNEP, *Analyzing the Resource Implication of Fisheries Subsidies: A Matrix Approach* (UNEP 2004).

9 Deepali Fernandes, "Running Into Troubled Waters – The Fish Trade and Some Implications?" *Evian Group Policy Brief* (November 2006), http://ssrn.com/abstract=1138271. p. 8.

10 J. S. Ferris and C. G. Plourde, "Labour Mobility, Seasonal Unemployment Insurance, and the Newfoundland Inshore Fishery," *Canadian Journal of Economics* 15(3) (1982): 426–41.

11 World Bank, "Module 6: Investment in Fisheries and Aquaculture," *Sourcebook: Agricultural Investment* (December 2006) http://go.worldbank.org/JRDC9SNHS0.

12 Council Regulation (EC) No 1198/2006 of 27 July 2006 on the European Fisheries Fund.

13 Elizabeth R. DeSombre, "Fishing Under Flags of Convenience," *Global Environmental Politics* 5(4) (November 2005): 73–94.

14 "Atlantic Dawn: Fishing For Trouble," *Ecologist* (April 2003): 18.

15 Figures are from the FAO Yearbook: Fishery and Aquaculture Statistics 2007, tables A-2 and A-3.

4 REGULATORY EFFORTS AND IMPACTS

1 Hawthorne Daniel and Francis Minot, *The Inexhaustible Sea* (New York: Dodd, Mead, 1954).

2   James M. Acheson, *The Lobster Gangs of Maine* (Hanover and London: University Press of New England, 1988).

3   Elinor Ostrom, *Governing the Commons: The Evolution of Institutions for Collective Action* (Cambridge: Cambridge University Press, 1990).

4   M. J. Peterson, "International Fisheries Management," in Peter M. Haas, Robert O. Keohane, and Marc A. Levy, eds, *Institutions for the Earth* (Cambridge: MIT Press, 1993), pp. 249–305.

5   "Proclamation No. 2668: Policy of the United States with Respect to Coastal Fisheries in Certain Areas of the High Seas" (28 September 1945), 10 *Fed. Reg.* 12304.

6   United Nations Convention on the Law of the Sea (1982), Article 56.

7   Lawrence C. Hamilton and Melissa J. Butler, "Outport Adaptations: Social Indicators through Newfoundland's Cod Crisis," *Human Ecology Review*, 8(2)(2001): 1–11.

8   Alanzo Aguilar Ibarra, Chris Reid, and Andy Thorpe, "Neo-liberalism and its Impact On Overfishing and Overcapitalization in the Marine Fisheries of Chile, Mexico, and Peru," *Food Policy*. 25(5) (2000): 599–622.

9   Vlad M. Kaczynski, and David L Fluharty, "European Policies in West Africa: Who Benefits From Fisheries Agreements?" *Marine Policy* 26 (2002): 75–93.

10 Ishaan Tharoor, "How Somalia's Fishermen Became Pirates," *Time* (18 April 2009), available online: http://www.time.com/time/world/article/0,8599,1892376,00.html.

11 Howard S. Schiffmann, *Marine Conservation Agreements: The Law and Policy of Reservations and Vetoes* (Leiden: Martinus Nijhoff Publishers, 2008).

12 Ebol Rojas, "Fisheries Observer Harassment and Interference – A Global Challenge," *APO Mail Buoy* 2(3)(Fall 2008): 6–11.

13 Michael J. Fogarty, and Steven A. Murawski, "Do Marine Protected Areas Really Work?" *Oceanus* (1 February 2005), http://www.whoi.edu/oceanus/viewArticle.do?id=3782.

14 IUCN, and World Commission on Protected Areas, "Protect Planet Ocean: Saving The World's Last Frontier," http://www.protectplanetocean.org/, date visited: 6 August 2010.

15 Christopher Costello, Steven D. Gaines, and John Lynham, "Can Catch Shares Prevent Fisheries Collapse?" *Science* 321 (19 September 2008): 1678–81.

16 Robin Allen, James Joseph, and Dale Squires, eds, *Conservation and Mangement of Transnational Tuna Fisheries* (Oxford: Blackwell, 2010).

5 AQUACULTURE

1 *FAO Annual Yearbook 2007: Fishery and Aquaculture Statistics* (Rome: FAO, 2009), p. xxii. Figures are by weight.
2 *FAO Annual Yearbook 2007: Fishery and Aquaculture Statistics* (Rome: FAO, 2009), p. xvi.
3 FAO, *The State of World Fisheries and Aquaculture* (Rome: FAO, 2009), pp. 17–18.
4 FAO, Fisheries and Aquaculture Department, "Main Cultured Species," http://www.fao.org/fishery/topic/13531/en.
5 *FAO Annual Yearbook 2007: Fishery and Aquaculture Statistics* (Rome: FAO, 2009), p. 28.
6 MICRA., "Asian Carp Threat to the Great Lakes," *River Crossings: The Newsletter of the Mississippi Interstate Cooperative Resource Association* 11(3)(2002): 1–2.
7 Laurie Jones, "Return of the Oysters," *Chow* (4 October 2006), http://www.chow.com/stories/10143.
8 Jennifer L. Molnar, Rebecca L. Gamboa, Carmen Revenga, and Mark D. Spalding, "Assessing the Global Threat of Invasive Species to Marine Biodiversity," *Frontiers in Ecology and the Environment* 6(9) (2008): 485–92.
9 *FAO Annual Yearbook 2007: Fishery and Aquaculture Statistics* (Rome: FAO, 2009), p. 28.
10 FAO Fisheries and Aquaculture Department, Cultured Aquatic Species Information Programme, Salmo salar, http://www.fao.org/fishery/culturedspecies/Salmo_salar/en.
11 Government of British Columbia, "Finfish Aquaculture – B.C. Salmon Aquaculture Industry," http://www.agf.gov.bc.ca/fisheries/bcsalmon_aqua.htm.
12 Jonathan R. Barton and Arnt Fløysand, "The Political Ecology of Chilean Salmon Aquaculture, 1982–2010: A Trajectory from Economic Development to Global Sustainability," *Global Environmental Change* (20 June 2010).
13 John Roach, "Sea Lice from Fish Farms May Wipe Out Wild Salmon," *National Geographic News* (13 December 2007), viewed

at http://news.nationalgeographic.com/news/2007/12/071213–
salmon-lice.html.

14 Alexei Barrionuevo, "Chile Takes Steps to Rehabilitate its
Lucrative Salmon Industry," *New York Times* (4 February 2009):
A6.

15 Paul Greenberg, "Tuna's End," *New York Times Magazine* (27
June 2010): 28–37, 44–8.

6 CONSUMERS AND CATCHES

1 Monterey Bay Aquarium Seafood Watch, "Developing
Sustainable Seafood Recommendations," http://www.
montereybayaquarium.org/cr/cr_seafoodwatch/content/media/
MBA_SeafoodWatch_RecommendationProcess.pdf (23 April
2008), date visited: 11 July 2010.

2 Environmental Defense Fund, "Health Alerts – Seafood
Selector," http://www.edf.org/page.cfm?tagID=17694, date
visited: 12 July 2010.

3 Carrie Brownstein, Mercédès Lee, and Carl Safina, "Harnessing
Consumer Power for Ocean Conservation," *Conservation
Magazine*, 4(4)(Fall 2003), available online at: http://www.
conservationmagazine.org/2008/07/harnessing-consumer-
power-for-ocean-conservation/.

4 Paul Ziobro, "Restaurants Mobilize to Save Fisheries," *Wall Street
Journal* (12 July 2010): B4.

5 Matt Reed, "Florida Restaurants Fight Off Fake Grouper," *USA
Today* (21 November 2006), available online at http://www.
usatoday.com/news/nation/2006-11-21-florida-fake-grouper_x.
htm.

Selected Readings

Anyone who wants to find out more information about global fishing and aquaculture practices should start with the United Nations Food and Agriculture Organization (FAO), which produces an overview volume every two years. The current version of this publication is FAO, *The State of World Fisheries and Aquaculture 2008* (Rome: FAO, 2009). The FAO is a fantastic source of information on fisheries, much of which is available on the web. A good starting point is the FAO Fisheries and Aquaculture Department, at http://www.fao.org/fishery.

Other books that provide an excellent overview of the development of the global fishing industry and its impact on the environment and on people include Callum Roberts, *The Unnatural History of the Sea* (Washington: Covelo Press, 2007); D. H. Cushing, *The Provident Sea* (Cambridge: Cambridge University Press, 1988); James R. McGoodwin, *Crisis in the World's Fisheries: People, Problems, and Policies* (Stanford: Stanford University Press, 1990); and Mark Kurlasky, *Cod: A Biography of the Fish that Changed The World* (New York: Walker and Company, 1997).

The story of Patagonian toothfish, discussed at some length in the introduction and conclusion to the book (and referenced elsewhere) is expanded upon in G. Bruce Knecht, *Hooked: Pirates, Poaching, and the Perfect Fish* (New York: Rodale Press, 2006). For further discussion of bluefin tuna, three good sources are Richard Ellis, *Tuna: A Love Story* (New York: Alfred A. Knopf, 2008); Douglas Whynott, *Giant Bluefin*

(New York: North Point Press, 1995); and Sasha Issenberg, *The Sushi Economy: Globalization and the Making of a Modern Delicacy* (New York: Gotham Books, 2007). Another engaging overview of some of the high-profile fish species – salmon, tuna, bass, and cod – in the ocean is Paul Greenberg, *Four Fish: The Future of the Last Wild Food* (New York: The Penguin Press, 2010).

Several chapters examine the background circumstances that underpin some of the problems facing fisheries. For information about the commons-type structure of fisheries resources, a good overview is provided in J. Samuel Barkin and George E. Shambaugh, eds, *Anarchy and the Environment: The International Relations of Common Pool Resources* (Albany: SUNY Press, 1999). A discussion of some ways local communities have been able to overcome the potential overuse of common pool resources can be found in Elinor Ostrom, *Governing the Commons: The Evolution of Institutions for Collective Action* (Cambridge: Cambridge University Press, 1990). More details about flags of convenience and the problems they pose for both fishing and global shipping generally can be found in Elizabeth R. DeSombre *Flagging Standards: Globalization and Environmental, Safety, and Labor Regulations at Sea* (Cambridge, MA: MIT Press, 2006). Two excellent attempts to account for the role of subsidies in the global fishing industry are Ussif Rashid Sumaila and Daniel Pauly, eds, *Catching More Bait: A Bottom-Up Re-Estimation of Global Fisheries Subsidies*, University of British Columbia, Fisheries Centre Research Reports 14(6) (2006) and UNEP, *Analyzing the Resource Implication of Fisheries Subsidies: A Matrix Approach* (UNEP 2004).

Chapter 4 addresses the most commonly used efforts to conserve fishery resources. The tale of local lobster conservation is engagingly examined in James M. Acheson, *The Lobster Gangs of Maine* (Hanover and London: University Press of

New England, 1988). An overview of fisheries management in the wake of widespread declaration by states of Exclusive Economic Zones can be found in Syma A. Ebben, Alf Håkon Hoel and Are K. Syndes, *A Sea Change: The Exclusive Economic Zone and Governance Institutions for Living Marine Resources* (Dordrecht: Springer, 2005). Further discussion of the role of regional fisheries management organizations in efforts to manage global fisheries can be found in Chapter 4 of Elizabeth R. DeSombre, *Global Environmental Institutions* (New York: Routledge, 2006); a particularly valuable illustration of the workings of the institution managing tuna and tuna-like species in the Atlantic Ocean is D. G. Webster, *Adaptive Governance: The Dynamics of Atlantic Fisheries Management* (Cambridge, MA: MIT Press, 2009).

Our optimism for the role of ITQs (or Catch Shares, as they are more frequently called in the United States) to change the success at fisheries management is reflected in Christopher Costello, Steven D. Gaines, and John Lynham, "Can Catch Shares Prevent Fisheries Collapse?" *Science* 321 (19 September 2008); pp. 1678–81. Further discussion of the role of marine protected areas in fishery conservation can be found at IUCN, and World Commission on Protected Areas, "Protect Planet Ocean: Saving The World's Last Frontier," http://www. protectplanetocean.org/; this website covers the latest developments in worldwide MPAs, and includes links to maps of where they are located around the world's seas.

Aquaculture is covered in the overview FAO publication mentioned above, and also in a specialized publication, FAO, *State of World Aquaculture* 2006 (FAO Fisheries Technical Paper No, 500)(Rome: FAO, 2006). That organization has the most thorough information on the state of aquaculture in the world on the portion of its website dedicated to aquaculture   (http://www.fao.org/fishery/aquaculture/en).   A study conducted under the auspices of FAO and other international organizations discusses the historical origins of aquaculture: Herminio R. Rabanal, "History of Aquaculture," ASEAN/UNDP/FAO Regional Small-Scale Coastal Fisheries Development Project (Manila, Philippines: FAO, 1988). Paul Molyneaux, *Swimming In Circles: Aquaculture and the End of Wild Oceans* (New York: Thunder's Mouth Press, 2007) provides an elaboration of the problems of aquaculture for global fish supply, from the perspective of someone engaged in several aspects of the fishing industry.

Those interested in finding out more about the usefulness of consumer strategies for protecting global fisheries should start with Carrie Brownstein, Mercédès Lee, and Carl Safina, "Harnessing Consumer Power for Ocean Conservation," *Conservation Magazine*, 4(4)(Fall 2003). Information about consumer options, including a variety of guides for which seafood is sustainable or safe to consume, is widely available on the web. The Monterey Bay Aquarium's Seafood Watch guides are available at http://www.montereybayaquarium.org/cr/seafoodwatch.aspx. The Environmental Defense Fund's calculator about how much of what fish species is safe to consume can be found at http://www.edf.org/page.cfm?tagID=17694. The list of fisheries certified by the Marine Stewardship Council, the primary sustainability certification organization for seafood, can be found at: http://www.msc.org.

Index